Acoustics Dictionary

ACOUSTICS DICTIONARY

QUADRILINGUAL

ENGLISH
GERMAN
FRENCH
DUTCH

Walter Reichardt

1983

SPRINGER-SCIENCE+BUSINESS MEDIA, B.V.

Library of Congress Cataloging in Publication Data
Reichardt, Walter
  Acoustics dictionary
  1. Sound—Dictionaries—Polyglot. 2. Dictionaries, Polyglot.
  I. Berends, E.-G. II. Rijnja, H.A.J. III. Title.
  QC221.5.B47   1982   534'03   82-14108
  ISBN 978-94-009-6789-2   ISBN 978-94-009-6787-8 (eBook)
  DOI 10.1007/978-94-009-6787-8

(c) 1983 Springer Science+Business Media Dordrecht
Originally published by Martinus Nijhoff in 1983
Softcover reprint of the hardcover 1st edition 1983

Preface

Interest in acoustics continues to increase. Although this branch of science was concerned primarily with the promotion of qualitative and quantitative sound transmission until a few decades ago, emphasis is currently placed also on the limitation of sound nuisance and, by extension, the setting of boundaries for permissible sound levels in places where people are found. This last aspect in particular is exercising more and more influence on the design of buildings and machines, and in town and country planning.

In addition, sound vibrations, because of their physical characteristics, are being used increasingly in disparate disciplines such as navigation, medical investigation and non-destructive materials research.

The flood of publications resulting from this increased interest in acoustics has led to a growing number of people being confronted with terminology which had until quite recently only been used by a relatively small group of specialists and had remained largely unknown as a result.

This four language dictionary, based on 'W. Reichardt, Technische Akustik; Berlin 1979', has been compiled to make not only this literature but also the nomenclature of equipment and instructions for their use accessible to the specialist and the interested layman.

As with any multilingual vocabulary, this dictionary also includes only a partial equivalent for the terms in the different languages in a number of places. In cases where it has proved impossible to find terms which are mutually equivalent, the compilers have used the English term as a standard and have in the first instance compared the terms in the other languages with the English term in question for points of correspondence.

The spelling of the Dutch term is based on the 'Woordenlijst van de Nederlandse taal' with the exception of a number of words beginning with 'geluid...'. On the basis of a resolution taken by the 'Nederlands Akoestisch Genootschap', all these words are written without the intermediary 's' including those words which do have an intermediary 's' in the 'Woordenlijst van de Nederlandse taal', e. g. 'geluidsfilm', 'geluidsgolf', 'geluidssterkte.'

Any comments or suggestions for improvement will be gratefully received by Kluwer Technische Boeken BV, P. O. Box 23, 7400 GA DEVENTER (The Netherlands).

Collaborators

E.-G. Berends (English and German)
Ing. F. Lahannier (French)
Ir. H. A. J. Rijnja (Dutch)

Directions for use

1. **The headwords have been placed in a strictly alphabetical order, as illustrated in the examples below**

distorted waveform
distortion factor
distortionless microphone
distortion of sound
distortion of sound field

sonics
sonic soldering
sound energy flux
sounding board
sound insulation

Magnetdraht
magnetischer Tonabnehmer
magnetisches Mikrofon
magnetische Tonaufzeichnung
Magnetkopfjustierung

Rauschanteil
rauschbedingte Wahrnehmungsgrenze
Rauschen
Rauschfaktor
Rauschspannung

amplificateur à coïncidence
amplificateur acoustique
amplificateur à large bande
amplificateur apériodique
amplificateur à résonance

son creux
sondage acoustique
sonde acoustique
son de transmission
sonde ultrasonique

akoestiek
akoestisch beeld
akoestische admittantie
akoestische roostertrilling
akoestisch inactief

absorberen
absorberende wand
absorberend materiaal
absorptie
absorptiecoëfficiënt

2. **Abbreviations and symbols used in this dictionary**

( )   nonstationary (nonsteady) noise = nonstationary noise *or* nonsteady noise
[ ]   spectrum [density] level = spectrum level *or* spectrum density level
/   Gleichklang/im = im Gleichklang
       sonorité pleine/de = de sonorité pleine
< >   these brackets contain explanations
s.   = see
s. a.   = see also
<sl>   = slang
<US>   American English

ENGLISH

ENGLISH

A

	ABC-test	s. A 3		
A 1	**ability to vibrate**	Schwingfähigkeit f	faculté f de vibrer	mogelijkheid om te trillen
A 2	**absence of noise,** freedom from noise, noiselessness	Rauschfreiheit f	absence f de trouble (bruit de fond, perturbations)	afwezigheid van ruis
A 3	**absolute bone conduction test,** ABC-test	absolute Knochenleitungsprüfung f	test m de la conduction osseuse absolue	absolute botgeleidingstest
A 4	**absolute pitch**	absolute Tonhöhe f <Musik>	tonalité f absolue	absolute toonhoogte
A 5	**absorb**	schlucken <Schall>	absorber	absorberen
	absorbent	s. A 7		
A 6	**absorber**	schallschluckende Anordnung f, Absorber m	absorbeur m de son	demper
A 7	**absorbing material,** absorbent	Schallschluckmaterial n, Schallschluckstoff m, Schluckstoff m	matériau m absorbant, absorbant m	absorberend materiaal n
A 8	**absorbing wall**	Schallschluckwand f, Schluckwand f	paroi f absorbante	absorberende wand
A 9	**absorption**	Absorption f	absorption f	absorptie
A 10	**absorption coefficient**	Schluckgrad m, Absorptionsgrad m	coefficient m d'absorption	absorptiecoëfficiënt
A 11	**absorption discontinuity**	Absorptionssprung m	discontinuité f (seuil m) d'absorption	sprong in de absorptie
A 12	**absorption loss**	Absorptionsdämpfungsmaß n	pertes fpl par absorption	absorptieverlies n
A 13	**absorption trap**	Absorptionssinke f	faille f (trou m) d'absorption	absorberende val
A 14	**absorptive surface**	Schluckfläche f	surface f absorbante	absorberend oppervlak n
A 15	**acceleration level**	Beschleunigungspegel m	niveau m d'accélération	versnellingspeil n
A 16	**acceleration pick-up**	Beschleunigungsempfänger m <Aufnehmer>	senseur m d'accélération	versnellingsopnemer
A 17	**acceleration transducer**	Beschleunigungsaufnehmer m	transducteur m d'accélération	versnellingstransducent
A 18/19	**accelerometer**	Beschleunigungsmesser m	accéléromètre m	accelerometer
	accent	s. A 21		f
A 20	**accent**	Betonung f, Hervorhebung f	accent m	accent
A 21	**accentuate,** accent	akzentuieren, verdeutlichen, betonen, hervorheben	accentuer	beklemtonen
A 22	**accentuation**	Akzentuierung f	accentuation f	beklemtoning
A 23	**accidental printing**	Kopiereffekt m	effet m de la copie accidentel	doordrukeffect n, echo-effect n
A 24	**accordion**	Akkordeon n	accordéon m	accordeon n
A 25	**accumulating stimulus**	einschleichender Reiz m, Einschleichreiz m	excitation f accumulative	accumulatieve prikkeling
A 26	**acetate disk**	Azetatschallplatte f	disque m acétate	acetaatplaat
	ackouometer	s. A 28		
A 27	**acoulalion**	Sprachrohr n <für Schwerhörigenunterricht>	cornet m acoustique pour l'éducation de sourds partiels	spreekhoorn
A 28	**acoumeter,** ackouometer, acoutometer	Akumeter n, Hörschärfemesser m	acoumètre m	gehoorscherptemeter
A 29	**acoumetric,** acoumometric	akumetrisch	acoumétrique	akoemetrisch
A 30	**acoumetry**	Akumetrie f, Hörschärfemessung f, Cochlearisprüfung f	acoumétrie f	akoemetrie
A 31	**acoumometric**	s. A 29		
	acoupedics	Akupädie f, Akupädiatrie f	acoupédie f	akoepedie
A 32	**acouphone**	s. H 38		
A 33	**acousia**	Hörsinn m	sens m auditif	gehoorzin
A 34	**acousma**	Hörhalluzination f	acousmie f	gehoorhallucinatie
	acoustic admittance	akustische Admittanz f, Flußadmittanz f, akustischer Leitwert m	admittance f acoustique	akoestische admittantie
	acoustical absorption coefficient	s. S 166		
A 35	**acoustical balance**	Hallbalance f	balance f acoustique	akoestische balans
A 36	**acoustical current**	Schallfluß m, akustischer Strom m	courant m (flux m) acoustique	akoestische flux
A 37	**acoustical homing device**	akustische Zielsuchvorrichtung f	appareil m de la localisation acoustique	akoestisch doelzoekend apparaat n
A 38	**acoustical insulation material**	Schallisolationsmaterial n	matériau m d'insonorisation	geluidisolerend materiaal n
A 39	**acoustical lattice vibration**	akustische Gitterschwingung f	vibration f acoustique de grille	akoestische roostertrilling
A 40	**acoustically inactive**	schalltot, akustisch unwirksam	acoustiquement neutre (inactif)	akoestisch inactief
A 41	**acoustically inert**	akustisch reaktionsträge	acoustiquement inert	akoestisch traag
A 42	**acoustical mass,** acoustic mass	akustische Masse f	masse f acoustique	akoestische massa
A 43	**acoustical propagation constant**	Schallausbreitungskonstante f	constante f acoustique de propagation	akoestische voortplantingsconstante
A 44	**acoustical reactance**	akustische Reaktanz f, Flußreaktanz f	réactance f acoustique	akoestische reactantie
A 45	**acoustical reciprocity theorem**	akustischer Reziprozitätssatz m	théorème m de réciprocité acoustique	theorema n van akoestische reciprociteit
A 46	**acoustical resistance**	akustische Resistanz f, Flußresistanz f, akustischer Widerstand m	résistance f acoustique	akoestische weerstand
A 47	**acoustical speech power**	Sprechleistung f	puissance f acoustique	akoestisch vermogen n van spraak
A 48	**acoustic altimeter**	akustischer Höhenmesser m	altimètre m acoustique	akoestische hoogtemeter
A 49	**acoustic amplifier**	akustischer Verstärker m	amplificateur m acoustique	akoestische versterker
A 50	**acoustic attenuation constant**	akustischer Dämpfungskoeffizient m	constante f d'atténuation acoustique	akoestische verzwakkingsfactor

A 51	acoustic auditorium design	akustische Gestaltung f von Vortragsräumen	aménagement m acoustique d'auditorium	akoestisch ontwerp n van gehoorzalen
A 52	acoustic baffle	Schallwand f	baffle m acoustique	akoestisch scherm n
A 53	acoustic beating	akustische Schwebungen fpl	battement m acoustique	akoestische zwevingen pl
A 54	acoustic board	Schalldämmplatte f	paroi f amortissante	geluiddempende plaat
A 55	acoustic buoy	Heulboje f	bouée f acoustique	brulboei
A 56	acoustic centre, effective acoustic centre	akustisches Zentrum n	centre m acoustique, centre acoustique effectif	akoestisch middelpunt n
A 57	acoustic chamber	schallabsorbierender Raum m	chambre f (enceinte f) acoustique	stille kamer
A 58	acoustic circuit element	akustisches Schaltelement n	élément m de circuit acoustique	akoestisch schakelelement n
A 59	acoustic clarifier	Klangreiniger m	épureur m de son	toonzuiveraar
A 60	acoustic compliance	akustische Nachgiebigkeit f	souplesse f acoustique	akoestische samendrukbaarheid
A 61	acoustic concentration	Schallbündelung f	concentration f acoustique	bundeling van geluid
A 62	acoustic consultant	akustischer Berater m	conseiller m acoustique	akoestisch adviseur
A 63	acoustic control acoustic control	Schallsteuerung f s. a. N 33	contrôle m acoustique	akoestische bediening
A 64	acoustic controller	akustischer Regler m, akustisches Stellglied n	régleur m (contrôleur m) acoustique	akoestische regelaar
A 65	acoustic corrector	Verzugsrechner m <Schallortung>	correcteur m acoustique (de délai)	looptijdcompensator
A 66	acoustic countermeasures	Lärmabwehrmaßnahmen fpl	mesures fpl de lutte contre le bruit	maatregelen pl tegen geluidhinder
A 67	acoustic damping	Schalldämpfung f	atténuation f sonore	geluiddemping
A 68	acoustic delay line	akustische Verzögerungsleitung f	ligne f de retard acoustique	akoestische vertragingslijn
A 69	acoustic depth finder, fathometer	Echolot n	sonde f [acoustique] à écho	echolood n
A 70	acoustic depth sounding, echo depth sounding, echo sounding	Echolotung f	sondage m acoustique, echo-sondage m	echopeiling
A 71	acoustic detecting apparatus	Horchgerät n	appareil m de détection acoustique	luisterapparaat n
A 72	acoustic detector	akustisches Ortungsgerät n	détecteur m acoustique	akoestische detector
A 73	acoustic dispersion	Dispersion f von Schallwellen, akustische Dispersion	dispersion f acoustique (d'ondes sonores)	akoestische dispersie
A 74	acoustic duct, auditory canal (meatus), acoustic meatus	Gehörgang m	canal m auditif	gehoorgang
A 75	acoustic efficiency	akustischer Wirkungsgrad m	rendement m acoustique	akoestisch rendement n
A 76	acoustic elasticity acoustic engineering	akustische Federung f s. A 268	élasticité f acoustique	akoestische slapheid
A 77	acoustic excitation	akustische Erregung f (Anregung f)	excitation f acoustique	akoestische aanstoting
A 78	acoustic feedback, howlround	akustische Rückkopplung f	réaction f acoustique	akoestische terugkoppeling
A 79	acoustic field impedance	Feldimpedanz f	impédance f de champ acoustique	akoestische impedantie
A 80	acoustic field resistance	Feldresistanz f	résistance f de champ acoustique	akoestische weerstand
A 81	acoustic figures	Schwingungsfiguren fpl	figures fpl de Chladni	trillingspatronen pl
A 82	acoustic flowmeter	Ultraschalldurchflußmesser m	fluxomètre m à ultra-sons	akoestische stromingsmeter
A 83	acoustic funnel	Schalltrichter m	cornet m (pavillon m) acoustique	geluidtrechter
A 84	acoustic grating	akustisches Gitter n	grille f acoustique	akoestisch rooster n
A 85	acoustic homing	akustische Zielansteuerung f	localisation f (orientation f) acoustique, recherche f acoustique de but	akoestische doelgeleiding
A 86	acoustician	Akustiker m, Toningenieur m	acousticien m, ingénieur m du son	akoesticus
A 87	acoustic image	akustisches Bild n	image f acoustique	akoestisch beeld n
A 88	acoustic impedance	akustische Impedanz f, Flußimpedanz f	impédance f acoustique (de flux)	akoestische impedantie
A 89	acoustic impedance tube	akustisches Impedanzrohr n	tube m d'impédance acoustique	geluidmeetpijp
A 90	acoustic inductance	akustische Induktivität f	inductance f acoustique	akoestische inductantie
A 91	acoustic inert (inertia)	akustische Trägheit f, Schallhärte f	inertie f acoustique	akoestische traagheid
A 92	acoustic inlet	Einsprache f, Einsprechöffnung f <des Mikrofons>	entrée f acoustique	geluidopening
A 93	acoustic insulation	Schalldämmung f	isolation f acoustique	akoestische isolatie
A 94	acoustic interactions	akustische Wechselwirkungen fpl	interactions fpl acoustiques	akoestische wisselwerkingen pl
A 95	acoustic interferometer	Schallinterferometer n	interféromètre m acoustique	akoestische interferometer
A 96	acoustic intrusion protection	akustischer Einbruchschutz m	alarme f acoustique anti-effraction	akoestische inbraakbeveiliging
A 97	acoustic leakage, acoustic scattering, scattering of sound	Schallstreuung f	pertes fpl acoustiques, dispersion f acoustique (sonore, du son)	geluidverstrooiing
A 98	acoustic lens acoustic mass acoustic meatus	akustische Linse f s. A 42 s. A 74	lentille f acoustique	akoestische lens
A 99	acoustic memory, acoustic storage	akustischer Speicher m, Ultraschallspeicher m	mémoire f acoustique	akoestisch geheugen n
A 100	acoustic mirage	akustische Spiegelung f	réflexion f acoustique	geluidspiegeling
A 101	acoustic mixer	Tonmischer m	mélangeur m de sons	geluidmengtafel
A 102	acoustic mobility	akustischer Mitgang m, Flußmitgang m	mobilité f acoustique	akoestische beweeglijkheid
A 103	acoustic mode	Schwingungsmode f	mode m acoustique (oscillatoire)	trillingswijze

A 104	**acoustic nerves**	Hörnerven *mpl*	nerf *m* acoustique	gehoorzenuwen *pl*
A 105	**acoustic non-specular reflection**	nichtspiegelnde akustische Reflexion *f*	réflexion *f* acoustique non spéculaire	niet-spiegelende geluidweer-kaatsing
A 106	**acousticon**	Hörgerät *n*	appareil *m* d'écoute	hoortoestel *n*
A 107	**acousticophobia**	Akustophobie *f*, Geräusch-angst *f*	acousticophobie *f*	lawaaivrees
A 108	**acoustic organ stop**	akustisches Orgelregister *n*	registre *m* acoustique d'orgue	akoestisch orgelregister *n*
A 109	**acoustic orientation**	akustische Ortung *f*, Schall-ortung *f*	orientation *f* acoustique	richtingzoeken met geluid
	acoustic oscillation	*s.* S 252		
A 110	**acoustic output**	Schalleistung *f*, akustische Ausgangsleistung *f*	puissance *f* acoustique de sortie	akoestisch uitgangsver-mogen *n*
A 111	**acoustic panel**	Schallwand *f*, Schallschirm *m*	panneau *m* acoustique	akoestisch paneel *n*
A 112	**acoustic pattern**	Klangbild *n*	image *f* acoustique	klankbeeld *n*
A 113	**acoustic perception, sound perception**	akustische Empfindung (Wahrnehmung) *f*, Schall-empfindung *f*, Schall-wahrnehmung *f*	perception *f* du son	akoestische perceptie
A 114	**acoustic phase constant**	akustischer Phasenkoeffi-zient *m* (Phasenwinkel *m*)	constante *f* (angle *m*) de phase acoustique	akoestische fasehoek
A 115	**acoustic plaster**	schallschluckender Putz *m*	revêtement *m* absorbant (insonorisant)	geluiddempende bekleding
A 116	**acoustic power**	Schalleistung *f*	puissance *f* acoustique	geluidvermogen *n*
A 117	**acoustic power concen-tration**	Schalleistungsbündelung *f*, Schallbündelungsgrad *m*	focalisation *f* de puissance sonore	bundeling van geluidvermo-gen
	acoustic pressure	*s.* S 222		
A 118	**acoustic pressure detector**	Schalldruckaufnehmer *m*	lecteur *m* de pression acoustique	geluiddrukopnemer
A 119	**acoustic probe**	Gehörsonde *f*, Hörprüfsonde	sonde *f* acoustique	geluidsonde
A 120	**acoustic propagation coefficient**	akustischer Ausbreitungs-koeffizient *m*	coefficient *m* de propagation acoustique	geluidvoortplantingscoëffi-ciënt
A 121	**acoustic properties of a room**	Hörsamkeit *f* eines Raumes	audibilité *f* d'une pièce (salle)	akoestische eigenschappen *pl* van een kamer
A 122	**acoustic pulse**	akustischer Impuls *m*	impulsion *f* acoustique	geluidstoot
A 123	**acoustic radiation pressure**	Schallstrahlungsdruck *m*	pression *f* de rayonnement acoustique	akoestische stralingsdruk
A 124	**acoustic radiator**	Schallstrahler *m*	diffuseur *m* (émetteur *m*) acoustique	geluidstraler
A 125	**acoustic radiometer**	Schallstrahlungsmesser *m*	radiomètre *m* acoustique	akoestische stralingsmeter
A 126	**acoustic ray, sound beam**	Schallstrahl *m*	rayon *m* acoustique	geluidbundel
A 127	**acoustic reception, sound reception**	Schallempfang *m*, Schallauf-nahme *f*	réception *f* du son	ontvangst van geluid
A 128	**acoustic reflection**	Schallreflexion *f*	réflexion *f* acoustique	weerkaatsing van geluid
A 129	**acoustic reflex**	akustischer Reflex *m*, reflek-torische Kontraktion *f* der Innenohrmuskeln	réflexe *m* acoustique de l'oreille interne	akoestische reflex
A 130	**acoustic refraction**	akustische Brechung *f*	réfraction *f* sonore	akoestische refractie
A 131	**acoustic regeneration**	Schallrückkopplung *f*	couplage *m* acoustique en retour, accrochage *m* acoustique	terugkoppeling van geluid
A 132	**acoustic reproduction**	Klangwiedergabe *f*, Schall-wiedergabe *f*	reproduction *f* sonore (acoustique)	geluidweergave
A 133	**acoustic resonator**	Schallresonator *m*	résonateur *m* acoustique	akoestische resonator
A 134	**acoustics**	Akustik *f*, Schalltechnik *f*, Tontechnik *f*	acoustique *f*, technique *f* du son	akoestiek
	acoustic scattering	*s.* A 97		
A 135	**acoustic shadow**	Schallschatten *m*	ombre *f* acoustique	geluidschaduw
A 136	**acoustic shock**	Knall *m*	choc *m* acoustique	knal
A 137	**acoustic short circuit**	akustischer Kurzschluß *m*	court-circuit *m* acoustique	akoestische kortsluiting
A 138	**acoustic slow motion**	akustische Zeitdehnung *f*, gedehnter Schallablauf *m*	étalement *m* de la propaga-tion, extension *f* du temps de parcours	vertraagde weergave
	acoustics of a room	*s.* R 184		
	acoustic source	*s.* S 244		
A 139	**acoustic stiffness** <obsolete>	akustische Steife *f*, Schall-härte *f* <veraltet>	inertie *f* acoustique	akoestische stijfheid
A 140	**acoustic stimulus**	Schallreiz *m*	excitation *f* acoustique	akoestische stimulus
	acoustic storage	*s.* A 99		
A 141	**acoustic streaming**	akustische Gleichströmung *f*	flux *m* acoustique continu	akoestische stroming
A 142	**acoustic stress**	Schalldruck *m*	pression *f* sonore	akoestische druk
A 143	**acoustic survey**	Schallüberwachung *f*	surveillance *f* acoustique	akoestische bewaking
A 144	**acoustic tile**	Schalldämmstein *m*	brique *f* absorbant le bruit	akoestische tegel
A 145	**acoustic transmission system**	Schallübertragungssystem *n*	système *m* de transmission acoustique	systeem *n* voor overdracht van geluid
A 146	**acoustic transmissivity**	Schalldurchlaßgrad *m*, Schallübertragungsgrad *m*	degré *m* de perméabilité acoustique	akoestische doorlaatbaarheid
A 147	**acoustic trauma**	akustisches Trauma *n*, Hör-verlust *m*	traumatisme *m* acoustique	akoestisch trauma *n*
	acoustic velocity	*s.* S 251		
A 148	**acoustic viscosimeter**	akustischer Viskositätsmesser *m*	viscosimètre *m* acoustique	akoestische viscosimeter
A 149	**acoustic viscous bound-ary layer**	akustische Zähigkeitsgrenz-schicht *f*	viscosité *f* acoustique des couches de séparation	akoestisch viskeuse grenslaag
A 150	**acoustic wave filter**	akustisches Filter *n*	filtre *m* acoustique	akoestisch golffilter *n*
A 151	**acoustimeter**	Geräuschmesser *m*	sonomètre *m*, acoustimètre *m*	geluidmeter
	acoutometer	*s.* A 28		
A 152	**action of acoustic resonance**	akustische Resonanz-wirkung *f*	effet *m* de résonance acoustique	werking van de akoestische resonantie
A 153	**action of ultrasounds**	Ultraschallwirkung *f*	effet *m* d'ultra-sons	werking van ultrageluid
A 154	**active sonar**	Aktiv-Sonar *n*	sonar *m* actif	actieve sonar
A 155	**active transducer**	aktiver Übertrager *m*	transducteur *m* actif	actieve transducent

A 156	**acuity meter,** acumeter	Hörschärfemeßgerät n	acuimètre m auditif	gehoorscherptemeter
A 157	**acuity of hearing,** auditory acuity acumeter	Hörschärfe f	acuité f auditive	gehoorscherpte
		s. A 156		
A 158	**acute**	schrill, akut	aigu	schril
A 159	**adaptation**	Anpassung f, Adaptierung f	adaptation f	adaptatie
A 160	**additional amplifier**	Zusatzverstärker m	amplificateur m additionnel	extra versterker
A 161	**additional stop**	Nebenregister n <Orgel>	registre m auxiliaire <d'orgue>	nevenregister n
A 162	**additional tones**	Zusatztöne mpl	tons mpl supplémentaires	bijtonen pl
A 163	**adjacent channel**	Nachbarkanal m	canal m adjacent (voisin, contigu)	aangrenzend kanaal n
A 164	**adjacent frequency**	Nachbarfrequenz f	fréquence f voisine (contiguë)	naastliggende frequentie
A 165	**adjustable equalizer**	stellbarer Entzerrer m	correcteur m de distorsion ajustable	regelbaar correctienetwerk n
A 166	**adjustment of contrast**	Dynamikregelung f	réglage m de contraste	contrastregeling
A 167	**admittance**	Admittanz f	admittance f	admittantie
A 168	**advancing wave,** progressive wave	fortschreitende Welle f, Vorwärtswelle f	onde f de propagation	lopende golf
A 169	**aeolian harp**	Äolsharfe f	harpe f éolienne, éolienne f	aeolusharp
A 170	**aerophone**	Aerophon n, Schalltrichter m	pavillon m acoustique	aërofoon
A 171	**air-bone gap,** bone-air gap	Hörschärfedifferenz f zwischen Luft- und Knochenleitung	différence f entre la conduction aérienne et la conduction osseuse	verschil n tussen lucht- en botgeleiding
A 172	**airborne insulation margin**	Luftschallschutzmaß n	degré m d'isolation contre les bruits aériens	isolatie van luchtgeluid
A 173	**airborne sound**	Luftschall m	son m se propageant dans l'air	luchtgeluid n
A 174	**airborne sound attenuation**	Luftschalldämpfung f	amortissement m du son aérien	verzwakking van luchtgeluid
A 175	**airborne sound source**	Luftschallquelle f	source f (émetteur m) de son aérien	luchtgeluidbron
A 176	**air chamber loudspeaker**	Luftkammer-Lautsprecher m	haut-parleur m à chambre pneumatique (de compression)	drukkamerluidspreker
A 177	**air conduction**	Luftleitung f	conduction f aérienne	geleiding van luchtgeluid
A 178	**air conduction receiver**	Luftleitungsempfänger m, Luftleitungsaufnehmer m <der Hörhilfe>	récepteur m par conduction aérienne	ontvanger voor luchtgeluid
A 179	**aircraft exposure level**	Flugzeugbelastungspegel m	niveau m de gêne par bruit d'avion	hinderpeil n voor vliegtuiglawaai
A 180	**air cushion**	Luftkissen n	coussin m d'air	luchtkussen n
A 181	**air damping**	Luftdämpfung f	amortissement m pneumatique	demping door luchtwrijving
A 182	**alien frequencies**	Fremdfrequenzen fpl, Fremdtöne mpl	fréquences fpl étrangères	vreemde frequenties pl
A 183	**align**	abgleichen	aligner	afregelen
A 184	**aliquot strings**	Aliquotsaiten fpl	cordes fpl aliquotes	aliquotsnaren pl
A 185	**aliquot tone**	Aliquotton m	son m aliquote	aliquottonen pl
A 186	**alloquism**	Sprechverständigung f, Sprachverständigung f, Sprachkommunikation f	compréhension f par le langage	spraakverstaanbaarheid
A 187	**all-pass amplifier**	Allpaßverstärker m	amplificateur m toutes fréquences	breedbandversterker
A 188	**all-pass filter section**	Allpaßfilterglied n	élément m de réseau compensateur de phase	allesdoorlatend filter n
A 189	**all-pass network**	Allpaß m	réseau m compensateur de phase	allesdoorlatend netwerk n
A 190	**altered chord,** chromatic chord	alterierter Akkord m	accord m altéré	gealtereerd akkoord n
A 191	**alto**	Alt m <Stimmlage>	alto m	alt
A 192	**alto clarinet,** basset-horn	Altklarinette f, Bassethorn n	clarinette f alto	altklarinet, bassethoorn
A 193	**alto-trombone**	Altposaune f	trombone m alto	altschuiftrompet, alttrombone
A 194	**ambient noise**	Umgebungsgeräusch n	bruit m ambiant	omgevingslawaai n
A 195	**ambiophony**	Ambiophonie f	ambiophonie f	ambiofonie
A 196	**amblyacousia,** hardness of hearing	Schwerhörigkeit f	amblyacousie f, dureté f d'oreille, surdité f	hardhorendheid
A 197	**amplifier frequency response**	Frequenzgang m des Verstärkers	plage f de réponse d'un amplificateur	frequentiekarakteristiek van de versterker
A 198	**amplifier noise**	Verstärkerrauschen n	bruit m de fond d'amplificateur	versterkerruis
A 199	**amplitude build-up**	⌐f Amplitudenaufschaukelung	croissance f d'amplitude	opslingering
A 200	**amplitude compressor**	Dynamikpresser m, Amplitudenpresser m	compresseur m d'amplitude	amplitude-compressor
A 201	**amplitude correction**	Amplitudenentzerrung f	correction f d'amplitude	amplitude-correctie
A 202	**amplitude discrimination**	Amplitudenselektion f	sélection f d'amplitude	amplitudeonderscheid n
A 203	**amplitude distortion**	Amplitudenverzerrung f	distorsion f d'amplitude	amplitudevervorming
A 204	**amplitude frequency characteristic**	Frequenzgang m	caractéristique f de fréquence en amplitude	amplitudefrequentiekarakteristiek
A 205	**amusia**	Musiktaubheit f, Unmusikalität f <absolut>	amusie f	amusikaliteit
A 206	**anacousia,** anacousis	Taubheit f <total>	surdité f totale	doofheid
A 207	**analyzer**	Analysiergerät n, Analysator m	analyseur m	analysator
A 208	**anechoic**	echofrei	libre d'écho	echoloos
A 209	**anechoic room**	schalltoter (reflexionsfreier) Raum m, Freifeldraum m	chambre f sourde (sans écho)	echoloze kamer
A 210	**anechoic studio**	Studio n mit kurzer Nachhallzeit, trockenes Studio n	studio m sans écho	galmvrije studio
A 211	**angle of rebound**	Abprallwinkel m	angle m de réflexion (rebondissement)	hoek van terugkaatsing

A 212	angular deviation loss	Richtungsdämpfungsmaß n	amortissement m (pertes fpl) de déviation angulaire	richtingsverlies n
A 213	angular directivity-factor	Richtungsfaktor m	facteur m de directivité angulaire	richtingsgevoeligheids-factor
	angular frequency	s. C 67		
A 214	annoyance	Lästigkeit f <eines Geräusches>	importunité f	geluidhinder
A 215	anti-distortion device	Entzerrer m	dispositif m anti-distorsion	vervormingscorrectieapparaat n
A 216	anti-hum capacitor	Entbrummkondensator m	condensateur m anti-ronfleur	ontbrommingscondensator
	antinode	s. L 83		
A 217	anti-noise	geräuschdämpfend	anti-bruit	geruisonderdrukkend
A 218	anti-noise microphone, noise-cancelling microphone	geräuschkompensierendes Mikrofon n	microphone m anti-bruité (à bruit compensé)	geruisonderdrukkende microfoon
A 219	antiresonance	Antiresonanz f	antirésonance f	antiresonantie
A 220	antiresonant circuit	Saugkreis m, Entkopplungskreis m	circuit m absorbant (anti-résonnant)	zuigkring
A 221	antiresonant coil	Entzerrungsdrossel f	inductivité f de correction	correctiespoel
A 222	anti-singing device	Einrichtung f zur Aufhebung der Pfeifneigung	dispositif m anti-sifflement	fluitfilter n
A 223	antiskating device	Antiskatingvorrichtung f <Phono>	disposition f antiskating	dwarsdrukcompensatie
A 224	aperiodic instrument	aperiodisch gedämpftes Instrument n	instrument m apériodique	aperiodisch gedempt instrument n
A 225	aphemesthesia	Worttaubheit f	aphémie f	woorddoofheid
A 226	apparent attenuation	Scheindämpfung f	atténuation f apparente	schijnbare verzwakking
A 227	appliance sound level	Armaturengeräuschpegel m <z. B. Wasserleitungen>	niveau m de bruit des installations	geruispeil n van een installatie
A 228	applied acoustics	angewandte Akustik f	acoustique f appliquée	toegepaste akoestiek
A 229	applied power	zugeführte Leistung f	puissance f appliquée (fournie)	geleverd vermogen n
A 230	applied shock	Anregungsstoß m, Stoßerregung f	choc m d'excitation	schokexcitatie
A 231	aquarium <sl>	Mischraum m	chambre f de mélange	mengkamer
A 232	architectural acoustics	Bau- und Raumakustik f	acoustique f architecturale	bouwakoestiek
A 233	archlute	Erzlaute f, Baßlaute f	luth m basse	basluit
A 234	arm resonance	Tonarmresonanz f	résonance f de bras (lecteur, pick-up)	armresonantie
A 235	arrival angle, incidence angle	Einfallswinkel m	angle m d'arrivée (d'incidence)	invalshoek
A 236	articulate	deutlich	articulé	gearticuleerd
A 237	articulation, intelligibility	Sprachverständlichkeit f	intelligibilité f	articulatie, verstaanbaarheid
	articulation of sentences	s. I 84		
A 238	articulation of words, discrete word intelligibility, intelligibility of words	Wortverständlichkeit f	intelligibilité f de mots (la parole)	woordverstaanbaarheid
A 239	articulation reduction	Minderung f der Verständlichkeit	réduction f de l'intelligibilité	vermindering van verstaanbaarheid
	artificial cranial bone	s. A 243		
A 240	artificial ear	künstliches Ohr n	oreille f artificielle	oorsimulator
A 241	artificial echo	künstliches Echo n	écho m artificiel	kunstmatige echo
A 242	artificial larynx	künstlicher Kehlkopf m	larynx m artificiel	strottehoofdsimulator
A 243	artificial mastoid, artificial cranial bone	künstliches Mastoid n	mastoïde m artificiel	rotsbeensimulator
A 244	artificial mouth	künstlicher Mund m	bouche f artificielle	mondsimulator
A 245	artificial voice	künstliche Sprache f	voix f artificielle	kunstmatige stem
A 246	assonance	Lautähnlichkeit f <besonders von Vokalen>	assonance f	assonantie
A 247	astatic microphone	Allrichtungsmikrofon n	microphone m multidirectionnel	ongerichte microfoon
A 248	atmospheric absorption	atmosphärische Absorption f, Luftabsorption f	absorption f atmosphérique	geluidabsorptie in de atmosfeer
A 249	atmospheric attenuation	atmosphärische Dämpfung f	atténuation f atmosphérique	geluidverzwakking in de atmosfeer
A 250	atmospheric noise, natural static	atmosphärisches Rauschen n	bruit m atmosphérique, parasites mpl atmosphériques	atmosferische ruis
A 251	atonal interval	atonales Interval n	intervalle m atonal	atonaal interval
A 252	attenuation characteristic	Dämpfungsverlauf m, Dämpfungsgang m	caractéristique f d'amortissement	verzwakkingskarakteristiek
A 253	attenuation coefficient, decay coefficient	Dämpfungskoeffizient m	coefficient m d'amortissement (d'atténuation)	verzwakkingscoëfficiënt
A 254	attenuation constant	Dämpfungskonstante f	constante f d'atténuation	dempingsconstante
A 255	attenuation correction	Dämpfungsentzerrung f	correction f d'amortissement	dempingscorrectie
A 256	attenuation pad, pad	Dämpfungsglied n	élément m atténuateur, affaiblisseur m, atténuateur m	verzwakker
A 257	audibility	Hörbarkeit f	audibilité f	hoorbaarheid
A 258	audible	hörbar, hörfrequent	audible	hoorbaar
A 259	audible Doppler enhancer	akustisches Doppler-Gerät n	appareil m acoustique à effet Doppler-Fizeau	omvormer voor hoorbare Dopplerfrequenties
A 260	audible frequency	Hörfrequenz f	fréquence f audible	hoorbare frequenties pl
A 261	audible region	Hörzone f	zone f d'audibilité	hoorbaar gebied n
A 262	audible sound	Hörschall m	son m audible	hoorbaar geluid n
A 263	audible spectrum	Hörschallspektrum n, Tonfrequenzspektrum n	spectre m des fréquences audibles	spectrum n van hoorbare frequenties
	audicle	s. H 38		
A 264	audience	Auditorium n	auditoire m	auditorium n
A 265	audio amplification	Niederfrequenzverstärkung f	amplification f basse fréquence	laagfrequente versterking

A 266	audio amplifier	Tonfrequenzverstärker m, Niederfrequenzverstärker m	amplificateur m de basse fréquence	laagfrequentversterker
A 267	audio channel	Tonfrequenzkanal m	canal m à fréquence audible	geluidkanaal n
A 268	audio engineering, acoustic engineering	Tontechnik f, NF-Technik f	technique f du son, technique des basses fréquences	geluidtechniek
A 269	audio feedback howl	Rückkopplungspfeifen n	sifflement m de superréaction	rondzingen
A 270	audio fidelity	Tonfrequenz-Wiedergabe-treue f	fidélité f de reproduction (réponse) acoustique	getrouwheid van weergave
A 271	audio frequency	Tonfrequenz f, Hörfrequenz f, Niederfrequenz f	fréquence f audible (sonore) basse fréquence	lage frequentie
A 272	audio frequency band	Tonfrequenzband n	bande f de (des) fréquences audibles	gebied n van de hoorbare frequenties
A 273	audio frequency band-pass filter	Niederfrequenzpaß m	filtre m passe-bas, filtre à basses fréquences	laagfrequent-banddoorlaat-filter n
A 274	audio frequency power	Tonfrequenzleistung f	puissance f basse fréquence	laagfrequent-vermogen n
A 275	audio frequency spectro-graph	Tonfrequenzspektrograf m	spectrographe m basse fréquence	laagfrequent-spectrograaf
A 276	audio frequency spectrometer	Tonfrequenzspektrometer n	spectromètre m basse fréquence	laagfrequent-spectrometer
A 277	audiogram	Audiogramm n, Schwell-wertkurve f, Hörverlust-kurve f	audiogramme m	audiogram n
	audiography	s. A 284		
A 278	audiologist	Audiologe m	audiologiste m	audioloog
A 279	audiology	Audiologie f	audiologie f	audiologie
A 280	audiometer, aurometer	Audiometer n	audiomètre m	audiometer
A 281	audiometric hearing loss	audiometrischer Hörverlust m	perte f audiométrique d'acuité auditive	audiometrisch gehoorver-lies n
A 282	audiometric room	Hörprüfraum m	espace m d'audiométrie	audiometrische kamer
A 283	audiometric technician	Audiometrist m	technicien m en audiométrie	audiometrisch technicus
A 284	audiometry, audiography	Audiometrie f	audiométrie f	audiometrie
A 285	audio monitoring	Kontrollabhören n	écoute f de contrôle	afluisteren
A 286	audio operator, sound supervisor, sound balancer ‹sl›	Tonmeister m	technicien m (opérateur m) du son	geluidtechnicus
A 287	audio oscillator	Tonfrequenzgenerator m, Tongenerator m	générateur m de fréquences acoustiques	toongenerator
A 288	audio peak limiter	Tonfrequenz-Amplituden-begrenzer m	limiteur m d'amplitude de fréquences acoustiques	laagfrequent-begrenzer
A 289	audiophile	HiFi-Enthusiast m, HiFi-Liebhaber m	audiophile m; amateur m de haute fidélité	geluidamateur
A 290	audio reception	Hörempfang m	réception f auditive	geluidontvangst
A 291	audio set-up	Phonoanlage f	ensemble m d'électrophone	geluidinstallatie
A 292	audio spectrum	Hörspektrum n	spectre m audible	spectrum n van hoorbare fre-quenties
A 293	audition	Hörvermögen n, Gehör n, Hörprobe f	audition f	gehoor n
A 294	auditor	Zuhörer m	auditeur m	toehoorder
A 295	auditorium	Zuhörerraum m	auditorium m, salle f de concert (spectacle)	gehoorzaal
A 296	auditorium noise, crowd noise, hall noise	Saalgeräusch n	bruit m d'ambiance (de salle)	zaalgeruis n
A 297	auditory	Zuhörerschaft f, Publikum n	auditoire m	publiek n
	auditory acuity	s. A 157		
	auditory canal	s. A 74		
A 298	auditory direction finding	Hörpeilung f	orientation f acoustique	richtinghoren
A 299	auditory localization	Ortung f nach Gehör	localisation f auditive	plaatsbepaling op het gehoor
	auditory meatus	s. A 74		
A 300	auditory nerve	Gehörnerv m	nerf m auditif	gehoorzenuw
A 301	auditory organ	Gehörorgan n	organe m du sens auditif	gehoororgaan n
A 302	auditory perspective	Hörperspektive f	perspective f auditorielle	gehoorperspectief n
A 303	auditory reflex	Hörreflex m	réflexion f sonore	gehoorreflex
A 304	auditory sensation	Schallwahrnehmung f	sensation f auditive	geluidwaarneming
A 305	auditory sensation area	Hörfläche f	aire f de sensation auditive	gehoorveld n
A 306	auditory sensitivity, hearing sensitivity	Hörempfindlichkeit f	sensibilité f auditive	oorgevoeligheid
A 307	auditory time error	akustischer Zeitfehler m	erreur f de temps auditive	subjectief tijdverschil n
A 308	auditory tube	Eustachsche Röhre f	trompe f d'Eustache	buis van Eustachius
A 309	aural	zum Ohr gehörig	auriculaire	op het oor betrekking heb-bend, oor . . .
A 310	aural check	Gehörprüfung f	examen m auriculaire	gehoortest
A 311	aural critical band	Frequenzgruppe f	bande f critique auriculaire (auditive)	kritische bandbreedte
A 312	aural dazzling	akustische Blendung f	éblouissement m acoustique	oorverdovend
A 313	aural harmonic	subjektive Harmonische f ‹im Ohr erzeugt oder empfunden›	harmonique f subjective ‹perçue subjectivement›	subjectieve harmonische
A 314	aural masking	Tonmaskierung f	masque m acoustique	maskering
A 315	aural-null direction finder	akustisches Peilgerät n	goniomètre m acoustique de zéro	richtingzoeker op het gehoor
A 316	aural resolving power	Auflösungsvermögen n des Ohres	pouvoir m de résolution de l'oreille	oplossend vermogen n van het oor
A 317	aural transducer	akustischer Wandler m	transducteur m acoustique	oortelefoon
A 318	auricle, auricula	Ohrmuschel f	pavillon m de l'oreille	oorschelp
A 319	auriform	ohrförmig	auriforme	oorvormig
A 320	auriscope	Otoskop n	otoscope m	otoscoop
A 321	aurist	Otologe m	otologiste m	otoloog
	aurometer	s. A 280		
A 322	auscultation	Auskultation f, Abhorchen n	auscultation f	auscultatie
A 323	auscultation tube	Stethoskop n	stéthoscope m	stethoscoop
A 324	auto-changer	automatischer Platten-wechsler m	changeur m automatique de disques	automatische platenwisselaar

A 325	auto fine tuning	automatische Feinabstimmung f	accord m automatique de précision	automatische fijn-afstemming
A 326	autoharp	Klaviaturzither f	cithare f à clavier	klavierciter
A 327	automatic bass compensation	automatische Baßanhebung f	compensation f automatique des basses	automatische compensatie van lage tonen
A 328	automatic volume contractor	Dynamikdränger m, Dynamikkompressor m	compresseur m de dynamique	dynamiekcompressor
A 329	automatic volume expander	Dynamikdehner m	expanseur m de dynamique	dynamiekvergroter
A 330	available noise power	verfügbare Rauschleistung f	puissance f de bruit disponible	beschikbaar ruisvermogen n
A 331	available power	maximal erhältliche Leistung f	puissance f maximale disponible	beschikbaar vermogen n
A 332	available power response	Schalleistung f ‹bei optimaler Anpassung›	puissance f de réponse disponible	frequentiekarakteristiek van het beschikbaar vermogen
A 333	average noise factor	Bandrauschzahl f	facteur m de bruit de bande	gemiddelde ruisfactor
A 334	average peak noise	mittlere Geräuschspitze f	amplitude f de pointe moyenne de bruit	gemiddelde piekwaarde van de ruis
A 335	average speech power	mittlere Sprechleistung f	puissance f acoustique moyenne	gemiddeld spraakvermogen n
A 336	axial response, axial sensitivity	Übertragungsfaktor m in Achsrichtung	sensibilité f (facteur m de réponse) axiale	gevoeligheid op de as
A 337	axial source level	Sonar-Sendepegel m	niveau m de puissance de source sonar	bronsterkte op de as ‹transducent›
A 338	axis of refraction	Brechungsachse f	axe m de réfraction	as van breking, brekingsas

B

B 1	babbling	Babbeln n, Murmeln n	murmure m	gekabbel n
	back coupling	s. F 19		
B 2	background loudspeaker	Hintergrundlautsprecher m	haut-parleur m de fond	decorluidspreker
B 3	background music, mood music	Hintergrundmusik f	coulisse f (musique f de fond) sonore	achtergrondmuziek
B 4	background noise	Systemeigengeräusch n, Grundgeräusch n	bruit m de fond propre	achtergrondgeruis n
B 5	background noise level	Grundgeräuschpegel m	niveau m de bruit de fond	achtergrondgeruisniveau n
B 6	background radiation	Grundstrahlung f	rayonnement m de fond	achtergrondstraling
B 7	backscattering	Rückstreuung f	dispersion f en retour	terugverstrooiing
B 8	backscattering cross section	Rückstreuquerschnitt m	section f de dispersion en retour	terugverstrooiingsdoorsnede
B 9	backscattering differential	Zielrückstreufaktor m	facteur m de réflexion différentielle	terugverstrooiingsfactor
B 10	backscattering strength, target strength	Zielrückstreumaß n	coefficient m de diffusion en retour, facteur m de réflexion de but	doelsterkte
B 11	backward masking	Rückverdeckung f	masquage m en retour	achterwaartse maskering
B 12	backward wave	Rückwärtswelle f	onde f en retour	teruglopende golf
	back wave	s. E 30		
B 13	baffle	Reflexionsplatte f, Schallwand f	paroi f réfléchissante	schermplaat
B 14	baffle blanket	Schallschluckbedeckung f, Schallschluckhülle f	revêtement m absorbant	absorptiehoes
B 15	baffle cloth	Bespannung f, Spannstoff m ‹des Lautsprechers›	étoffe f de haut-parleur	luidsprekerdoek n
B 16	bagpipes	Dudelsack m	cornemuse f, musette f	doedelzak
B 17	balance	Gleichgewicht n, Ausgewogenheit f ‹von Instrumentengruppen›	balance f, équilibre m	evenwicht n
B 18	balance	Abgleich m, Symmetrie f, Kompensation f, Ausgleich m, Gleichgewicht n	balance f, équilibre m, symétrie f	balans
B 19	balanced	symmetriert, angepaßt, abgeglichen	symétrisé, accordé	evenwichtig
B 20	balanced amplifier, push-pull amplifier	Gegentaktverstärker m	amplificateur m symétrique (de balance, push-pull)	balansversterker
B 21	balanced armature	Vierpolankersystem n	attelage m quadripolaire	vierpolig magneetsysteem n
B 22	balanced armature loudspeaker	symmetrischer vierpoliger magnetischer Lautsprecher m	haut-parleur m à système symétrique quadripolaire	luidspreker met vierpolig magneetsysteem
B 23	balanced armature pick-up system	symmetrisches Tonabnehmersystem n	pick-up m (lecteur m) à système symétrique	pick-up met vierpolig magneetsysteem, symmetrische pick-up
B 24	balanced condition	Gleichgewichtszustand m	état m d'équilibre	uitgebalanceerd
B 25	balanced detector	Frequenzwaagendemodulator m	détecteur m à balance de fréquences	symmetrische detector
B 26	balanced modulator	Gegentaktmodulator m	modulateur m symétrique	ringmodulator
B 27	balanced output	symmetrischer Ausgang m	sortie f symétrique	symmetrische uitgang
B 28	balanced quadripole	symmetrischer Vierpol m, symmetrisches Zweitor n	quadripôle m symétrique	symmetrische vierpool
B 29	balanced transformer	symmetrischer Übertrager m	transformateur m symétrique	symmetrische transformator
B 30	balance of moments	Momentengleichgewicht n	équilibre m des moments	evenwicht n in momenten
B 31	balance-to-unbalance transformer	Symmetrieübertrager m	transformateur m symétrique	symmetreertransformator
B 32	balancing loop	Symmetrierschleife f	boucle f de symétrisation	symmetreerlus
B 33	balancing method	Nullabgleichmethode f	accord m de symétrie par tarage sur zéro	nulmethode
B 34	Baldwin's electronic organ	Baldwin-Orgel f	orgue m électronique de Baldwin	Baldwin-orgel n
B 35	bamboo pipe	Bambusflöte f	flûte f de bambou	bamboefluit
B 36	band	Band n, Frequenzband n	bande f de fréquence	frequentieband
B 37	band edge	Bandgrenze f	limite f (bord m) de bande	bandgrens

B 38	band elimination filter	Bandsperrfilter n	filtre m de bande éliminatoire	bandsperfilter n
B 39	band merit	Bandgüte f	qualité f de bande	kwaliteitsfactor van de band
B 40	band-pass filter	Bandpaßfilter n	filtre m de bande	banddoorlaatfilter n
B 41	band pressure level	Bandschalldruckpegel m, Schalldruckpegel m in einem Frequenzband	niveau m de pression sonore dans une bande	geluiddrukpeil n in een frequentieband
B 42	band-rejection filter, band-stop filter	Bandsperre f, Bandsperrfilter n	filtre m coupe (d'élimination de) bande	bandsperfilter n
B 43	band relay	Übertragungsanlage f <für Musik-Band>	équipement m de retransmission	bandtransformatie
B 44	band space	Bandabstand m	intervalle m de bande	bandafstand
B 45	band spread	Banddehnung f, Bandspreizung f	étalement m de bande	bandspreiding
	band-stop filter	s. B 42		
B 46	band switch	Bereichsschalter m, Wellenumschalter m, Kanalschalter m	commutateur m de bande	golfbandschakelaar
B 47	bandwidth	Bandbreite f	largeur f de bande	bandbreedte
B 48	bandwidth control	Trennschärferegelung f, Bandbreiteregelung f	contrôle m de la sélectivité	bandbreedteregeling
B 49	bang	[scharfer] Knall m	bang m	knal
	bank of keys	s. K 5		
B 50	baritone, barytone	Bariton m	bariton m	bariton <stem>
B 51	baritone, euphonium	Baritonhorn n	bariton m	bariton
B 52	Barkhausen phon	Barkhausen-Phon n	phone m Barkhausen	barkhausen-foon
B 53	barrel organ, street organ	Drehorgel f	orgue m de Barbarie, orgue mécanique Limonaire	draaiorgel n
	barytone	s. B 50		
B 54	base band amplifier	Basisbandverstärker m	amplificateur m de base	basisversterker
B 55	basic frequency	Hauptfrequenz f, Grundfrequenz f	fréquence f principale	grondfrequentie
B 56	basic noise criterion	Grundlärmkriterium n	critère m de bruit de fond	fundamenteel geruiscriterium n
B 57	basilar membrane <of the ear>	Basilarmembran f <des Ohres>	membrane f basilaire <dans l'oreille interne>	trommelvlies n
B 58	bass, basso	Baß m, Baßstimme f	basse f, voix f de basse	bas
B 59	bass-baritone	Baßbariton m	basse-bariton f	bas-bariton
B 60	bass boost, bass lift (emphasis)	Tiefenanhebung f, Baßanhebung f	renforcement m des basses	versterking van de lage tonen
B 61	bass clarinet	Baßklarinette f	clarinette-basse f	basklarinet
B 62	bass compensation	Baßentzerrung f	correction f des basses	correctie voor lage tonen
B 63	bass control	Baßregelung f	contrôle m des basses	lagetonenregeling
B 64	bass-cut filter	Baßfilter n	filtre m suppresseur de basses	basfilter n
B 65	bass drum	Pauke f	grosse caisse f	pauk
	bass emphasis	s. B 60		
	basset-horn	s. A 192		
	bass lift	s. B 60		
B 66	bass lift and cut	Baßanhebung f und -beschneidung f	renforcement m et coupure f des basses	versterking en verzwakking van lage tonen
B 67	bass loudspeaker	Tieftonlautsprecher m, Baßlautsprecher m	haut-parleur m de basses	lagetonenluidspreker
B 68	bass masking	Baßabdeckung f	couverture f des basses	basmaskering
B 69	bass notes	Baßtöne mpl	notes fpl basses	lage tonen pl
	basso	s. B 58		
	basso continuo	s. F 26		
B 70	bassoon, curtal	Fagott n	basson m	fagot
B 71	basso profundo	tiefer Baß m	basse f profonde, contre-basse f	contrabas
B 72	bass reflex cabinet	Baßreflexgehäuse n	enceinte f basse-reflex	basreflexkast
B 73	bass response	Baßwiedergabe f	réponse f des basses	lagetonenweergave
B 74	bass singer	Bassist m	basse m <chanteur>	bas
B 75	bass string	G-Saite f	corde f de sol	g-snaar
B 76	bass trombone	Baßposaune f	trombone-basse m	bastrombone
B 77	bass viol	Gambe f	viole f de gambe	viola de gamba, beenviool
B 78	bass violin	Cello n	violoncelle m	violoncel
B 79	bathythermogram	Bathythermogramm n	bathythermogramme m	bathythermogram n
B 80	bazooka	Sperrtopf m	adaptateur m symétrie-asymétrie	symmetreertransformator
	BC	s. B 154		
B 81	beaming	Bündelung f	concentration f, focalisation f	bundeling
B 82	beam of sound	Schallstrahlbündel n	faisceau m sonore	geluidbundel
	beam pattern	s. D 123		
B 83	beam width	Strahlbreite f	largeur f de faisceau	bundelbreedte
B 84	beat	schlagen, klopfen <Takt>	battre la mesure	de maat slaan
B 85	beat	schweben	battre, interférer	zweven
B 86	beat	Takt m	mesure f	maat
B 87	beat	Schwebung f, Flatterschwingung f	battement m, interférence f	zweving
B 88	beat amplitude	Schwebungsamplitude f	amplitude f de battement	zwevingsamplitude
B 89	beat cycle	Schwebungsperiode f	période f de battement	zwevingsperiode
B 90	beater	Schlegel m <für Pauke>	mailloche f de grosse caisse	trommelstok
B 91	beat frequency	Schwebungsfrequenz f	fréquence f de battement	zwevingsfrequentie
B 92	beat indicator, metronome	Taktgeber m, Metronom n	métronome m	metronoom
B 93	beating	Schwebungsvorgang m	battement m	het zweven n
B 94	beat note	Schwebungston m	note f de battement	zwevingstoon
B 95	beat note distortion	Schwebungsverzerrung f	distorsion f de battement	zwevingsvervorming
B 96	beat note pitch	Schwebungstonhöhe f	hauteur f de la note de battement	zwevingstoonhoogte
B 97	beat the time	den Takt schlagen	battre la mesure	de maat slaan
B 98	beep	Piepton m	sifflement m	bandsnelheidssignaal n

B 99	behavioral audiometry	Verhaltensaudiometrie f, Spielaudiometrie f	audiométrie f de comportement	gedragsaudiometrie
B 100	Bekesy audiometer	halbautomatisches Audiometer n	audiomètre m semi-automatique	halfautomatische audiometer
B 101	bel	Bel n	bel m	bel
B 102	bell	glockenförmiger Schalltrichter m	pavillon m acoustique en forme de cloche	beker
B 103	bell	Glocke f	cloche f	klok
B 104	bell	Klingel f	clochette f, grelot m	bel
B 105	bell lyra, chinese pavilion, jingling Johnny <sl>, turkish crescent	Schellenbaum m	lyre f d'harmonie, chapeau m chinois	schellenboom
B 106	bell opening	Schalltrichteröffnung f	ouverture f de pavillon	trechteropening
B 107	bellow	brüllen, laut schreien	crier très haut	brullen
B 108	bellow	Gebrüll n, Geschrei n	rugissement m, cris mpl puissants	gebrul n
B 109	bellows	Balg m	soufflet m	balg
B 110	bellows stop	Balgregister n	registre m de soufflet primaire	balgregister n
B 111	bender transducer	piezoelektrischer Übertrager m	transducteur m piézo-électrique	transducent met buigtriller, piëzo-elektrische transducent
B 112	bending wave	Biegewelle f	onde f de flexion	buiggolf
B 113	Beranek scale	Beranek-Skala f	échelle f de Beranek	schaal van Beranek
B 114	best audible pressure	optimaler Schalldruck m	pression f sonore pour l'audition optimale	geluiddruk voor optimale verstaanbaarheid
B 115	bias out	durch Gegenkopplung kompensieren	compenser par contre-réaction	compenseren door tegenkoppeling
	biaural	$s.$ B 125		
B 116	biauricular	beide Ohren betreffend	biauriculaire	voor beide oren
B 117	bidirectional	zweiseitig (beidseitig) gerichtet	bidirectionnel	tweezijdig gevoelig
B 118	bidirectional microphone	Zweirichtungsmikrofon n	microphone m bidirectionnel	tweezijdig gevoelige microfoon
B 119	bidirectional transducer	Zweirichtungsübertrager m	transducteur m bidirectionnel	tweezijdig gevoelige transducent
B 120	bilabial	bilabial, mit beiden Lippen gesprochen	bilabial	bilabiaal
B 121	bilateral characteristic	zweiseitige (achtförmige) Richtcharakteristik f	caractéristique f en huit	achtvormig richtingsdiagram n
B 122	bilateral hearing aid	beidohrige Hörhilfe f <mit gemeinsamem Verstärker>	aide f acoustique bilatérale	bilateraal hoortoestel n
B 123	bilaterally matched two-port	bilateral angepaßtes Zweitor n	quadripôle m à adaptation symétrique	aan weerskanten aangepaste vierpool
B 124	bilateral microphone	Achtermikrofon n	microphone m à caractéristique en huit	bilaterale microfoon
B 125	binaural, biaural, binotic	beidohrig	biaural	twee-orig
B 126	binaural effect	Raumtoneffekt m	effet m bi-auriculaire	ruimtelijk horen
B 127	binaural hearing, stereophonic hearing	zweiohriges (räumliches) Hören n	audititon f bi-auriculaire, écoute f biaurale	horen met twee oren
B 128	binaural hearing aid	beidohrige Hörhilfe f <mit getrennten Verstärkersystemen>	aide f acoustique biaurale	binauraal hoortoestel n
B 129	binaural level difference, binaural ratio	Schallpegeldifferenz f zwischen beiden Ohren	différence f binaurale de niveau sonore	niveauverschil n tussen beide oren
B 130	binaural recorder	stereofoner Tonschreiber m	enregistreur m stéréophonique	stereofonische opnemer
B 131	binaural sound detection	beidohrige (räumliche) Schallortung f	orientation f bi-auriculaire	geluiddetectie met beide oren
B 132	bind	Bindebogen m <Musik>	liaison f	verbinding
	binotic	$s.$ B 125		
B 133	birdies	Störungen fpl durch hohe Töne, Zwitschern n	parasitage m par sifflements, gazouillis m	getjilp n
B 134	blank channel	unbesetzter (freier) Kanal m	canal m libre	vrij kanaal n
B 135	blasting	Übersteuerung f <des Mikrofons>, Dröhnen n <bei Funkempfang>	surcharge f, saturation f du microphone	overbelasting van de microfoon
B 136	blend	verschmelzen, vermischen	mélanger	vermengen
B 137	blending of sounds	Verschmelzen n von Klängen	mélange m de sons	vermenging van geluiden
B 138	block	zustopfen <Empfänger>, festbremsen <Membran>	bloquer	blokkeren
B 139	blocked diaphragm pressure	Druck m bei blockierter (festgebremster) Membran	pression f avec membrane bloquée	druk bij geblokkeerd membraan
B 140	blocked impedance	elektrische Eingangsimpedanz f bei blockiertem Schwingungssystem, Block-Impedanz f	impédance f d'entrée avec attelage bloqué	geblokkeerde impedantie
B 141	block flute	Blockflöte f	flûte f sans clés, pipeau m	blokfluit
B 142	blocking attenuation	Echosperrdämpfung f	affaiblissement m d'écho	blokkeerverzwakking
B 143	blocking characteristic	Sperrcharakteristik f	caractéristique f de blocage	blokkeerkarakteristiek
B 144	bloomy	dumpf <Ton>	sourd <son>	dof
B 145	bloop <sl>	Klebestellengeräusch n <Magnetband>	bruit m de raccordement collé	lasgeruis n
B 146	blooper	Empfänger m, der Störsignale sendet	récepteur m produisant des parasites	stoorsignalen uitzendende ontvanger
B 147	blow pressure	Anblasdruck m <bei Lippenpfeifen>	pression f de souffle	aanblaasdruk
B 148	blur	verwischen	estomper	wazig
B 149	blur circle	Unschärfering m	cercle m de flou	onscherptekring
B 150	blur level	Klirrpegel m	niveau m de distorsion	vervormingspeil n
B 151	blurred voice	verzerrte (unklare) Sprache f	voix f déformée	vervormde spraak
B 152	boiler maker's deafness	Kesselschmiedtaubheit f, berufsbedingte Taubheit f	surdité f des chaudronniers	ketelmakersdoofheid

2*

B 153	bombardon bone-air gap	Bombardon n s. A 171	contrebasse f <à vent>	bombardon
B 154	bone conduction, BC	Knochenleitung f	conduction f osseuse	botgeleiding
B 155	bone-conduction head-phone, bone-conduction receiver	Knochenleitungshörer m	récepteur (écouteur) m à conduction osseuse	botgeleidingstelefoon
B 156	bone-conduction micro-phone bone-conduction receiver	Knochenleitungsmikrofon n s. B 155	microphone m à conduction osseuse	botgeleidingsmicrofoon
B 157	bone-conduction vibra-tor	Knochenschallerzeuger m	vibrateur m pour conduction osseuse	botgeleidingstriller
B 158	boom	Dröhnen n	grondement m	bulderen
B 159	boomer, low-frequency loudspeaker, woofer	Tieftonlautsprecher m	haut-parleur m de basses (graves)	lagetonenluidspreker
B 160	boominess	[hohles] Dröhnen n des Mikrofons	grondement m de micro-phone	gebulder n
B 161	boom microphone	Galgenmikrofon n, Schwenkarmmikrofon n	microphone m à potence	hengelmicrofoon
B 162	boom pram	Ständer m des Mikrofon-galgens	support m (pied m) de microphone	hengelstandaard
B 163	boomy	[tief] dröhnend <Instru-ment>	bas, grondant	bulderend
B 164	boomy sound	hohler Ton m	son m creux	bulderend geluid n
B 165	bottom scattering coefficient	Oberflächenstreukoeffizient m	coefficient m de dispersion de surface	verstrooiingscoëfficiënt van de bodem
B 166	bottom scattering strength	Oberflächenrückstreumaß n	coefficient m de dispersion d'une surface réfléchis-sante	terugverstrooiingsfactor van de bodem
B 167	boundary layer approach	Grenzschichtnähe f	proximité f d'une couche de surface	benadering van een grens-laag
B 168	bow	Bogen m <Streichinstru-mente>	archet m	strijkstok
B 169	bow stroke	Bogenstrich m	trait m d'archet	streek
B 170	branch resistance	Zweigresistanz f	résistance f de branche	aftakkingsweerstand
B 171	brass band	Blasorchester n	orchestre m à vent	blaasorkest n
B 172	breast transmitter	Brustmikrofon n	microphone m plastron (à collier)	omhangmicrofoon met zender
B 173	breathing microphone	atmendes Mikrofon n	respiration f du microphone	ademende microfoon
B 174	breathing of micro-phone	Atmen n des Mikrofons	respiration f du microphone	het ademen n van een microfoon
B 175	bridge arm	Brückenzweig m	branche f de pont	brugtak
B 176	bridge balance bridge network	Brückenabgleich m s. L 13	ajustage m de pont (zéro)	brug-evenwicht n
B 177	brightness	Helligkeit f <des Tones>	clarté f du son	helderheid
B 178	brilliance	helle Klangfarbe f, glänzende, schillernde Tönung f	brillance f	helderheid
B 179	broadband amplifier, wand band amplifier	Breitbandverstärker m, ⌐m aperiodischer Verstärker	amplificateur m apériodique (à large bande)	breedbandige versterker
B 180	broadband balun	Breitbandsymmetrieüber-trager m	transformateur m symétri-que à large bande	breedbandige symmetrische transformator
B 181	broadband filter	Filter n großer Durchlaß-breite, breitbandiges Filter	filtre m à grande bande	breedbandig filter n
B 182	broadband noise	Breitbandgeräusch n	bruits mpl de bande large	breedbandige ruis
B 183	broadcasting centre	Funkhaus n	centre m d'émission radio-phonique	omroepcentrum n
B 184	broadcast interference	Rundfunkstörung f	parasites mpl radiophoniques	radiostoring
B 185	broadcast listener, listener-in	Rundfunkhörer m	auditeur m [de radio]	luisteraar
B 186	broadly tuned	unscharf abgestimmt	mal accordé	onscherp afgestemd
B 187	broad tuning	unscharfe Abstimmung f	accord m flou	onscherpe afstemming
B 188	bubbling noise	Blasengeräusch n, Blubber-geräusch n, brodelndes Geräusch n	bruits mpl de bouillonne-ment	geborrel n
B 189	Buchmann and Meyer pattern, Christmas-tree pattern wax picture	Buchmann- und Meyer-Bild n	image f de Buchmann [et Meyer]	Buchmann-Meyerpatroon n
B 190	buffer amplifier	Trennverstärker m	amplificateur m tampon, étage m adaptateur d'impédance	bufferversterker
B 191	bugle	Bügelhorn n	bugle m	signaalhoorn
B 192	bugle sound	Hörnerklang m	son m de cornets	hoornsignaal n
B 193	building-up	Aufschaukelung f	amorçage m d'oscillations	opslingering
B 194	building-up time	Einschwingzeit f, Auf-schaukelzeit f	période f d'amorçage, temps m d'amorçage	opslingeringstijd
B 195	bullet amplifier	Vorverstärker m	préamplificateur m	voorversterker
B 196	bull horn	Hochleistungslautsprecher m mit Richtcharakteristik	haut-parleur m directionnel à grande puissance	hoornluidspreker
B 197	bull roarer	Schwirrholz n	planchette f ronflante	bromhout n
B 198	bungalow <sl>	schalldichte Zelle f <Film-aufnahmestudio>	studio m insonorisé, cellule f acoustiquement isolée	geluiddichte cabine
B 199	burst	Impuls m, Stoß m	impulsion f, choc m	stoot
B 200	button	Kapsel f <des Mikrofons>	capsule f	knoopmicrofoon
B 201	button transmitter	Kapselmikrofon n	microphone m à capsule	knoopmicrofoon met zender
B 202	buzz	summen	bourdonner	zoemen
B 203	buzzer	Magnetsummer m	buzzer m	zoemer
B 204	buzzer wavemeter	Summerwellenmesser m	ondemètre m à buzzer	golfmeter met zoemer
B 205	buzzing noise	Summgeräusch n	bruits mpl de bourdonne-ment	zoemend geruis n
B 206	buzzing sound	Summton m	bourdonnement m	zoemend geluid n
B 207	by-pass transmission	Nebenwegübertragung f	transmission f par dérivation	neventransmissie

C

C 1	**calibration**	Kalibrierung *f*, Abgleich *m*	calibrage *m*, ajustage *m*, étalonnage *m*	kalibratie
C 2	**canal resonance**	Gehörgangsresonanz *f*	résonance *f* du canal auriculaire	gehoorgangsresonantie
C 3	**canaries**	Zwitschern *n*	gazouillis *m*	getjilp *n*
C 4	**canned music** <sl>	Konservenmusik *f* <sl>	musique *f* de conserve <sl>	ingeblikte muziek
C 5	**canning** <sl>	Tonaufzeichnung *f* <sl>	enregistrement *m* musical	geluidopname
C 6	**cap**	Ohrmuschel *f* <des Hörers>	capsule *f* <de l'écouteur>	oorkap
C 7	**capacitive pick-up**	kapazitiver Tonabnehmer *m*	lecteur *m* statique (capacitif)	capacitieve pick-up
C 8	**capacitor loudspeaker, condenser loudspeaker**	Kondensatorlautsprecher *m*, elektrostatischer Lautsprecher *m*	haut-parleur *m* électrostatique (statique)	condensatorluidspreker
C 9	**capacitor microphone, condenser microphone, electrostatic microphone**	Kondensatormikrofon *n*	microphone *m* statique (électrostatique)	condensatormicrofoon
C 10	**capacitor receiver**	Kondensatorkopfhörer *m*	écouteur *m* statique	condensatortelefoon
C 11	**caption buzz**	Schriftbrumm *m* <Fernsehen>	ronflement *m* <télévision pendant la transmission d'écriture>	brommen tijdens ondertiteling
C 12	**carbon membrane**	Kohlemembran *f*	membrane *f* en charbon	koolmembraan
C 13	**carbon microphone, granule microphone**	Kohlemikrofon *n*	microphone *m* à charbon	koolmicrofoon
C 14	**carbon noise**	Kohlekörnergeräusch *n*	bruits *mpl* de grenaille	koolkorrelgeruis *n*
C 15	**cardioid microphone, microphone with cardioid characteristic**	Mikrofon *n* mit Nierencharakteristik, Nierenmikrofon *n*	microphone *m* à [caractéristique] cardioïde	cardioïdemicrofoon
C 16	**carillon**	Glockenspiel *n*, Turmglockenspiel *n*; Glockenspielmusik *f*; Stahlspiel *n* <Orgelregister>	carillon *m*	carillon *n*
C 17	**carrier noise**	Trägerrauschen *n*	bruit *m* de porteuse	draaggolfruis
C 18/9	**carrier signal**	Trägersignal *n*	signal *m* porteur	draaggolfsignaal *n*
C 20	**cart-type mixer**	fahrbares Mischpult *n* <Studiotechnik>	pupitre *m* mélangeur mobile	verrijdbare mengtafel
C 21	**case noise** <of hearing aid>	Reibungsgeräusch *n* <Hörhilfe>	bruits *mpl* de frottement <des appareils de prothèse auditive>	kledinggeruis *n*
C 22	**castanet**	Kastagnette *f*	castagnettes *fpl*	castagnetten *pl*
C 23	**cathodophone, glow discharge microphone**	Katodofon *n*	cathodophone *m*	kathodofoon
C 24	**cavitation**	Kavitation *f*	cavitation *f*	cavitatie
C 25	**cavitation bubble**	Kavitationsblase *f*	bulle *f* de cavitation	cavitatiebel
C 26	**cavitation noise**	Kavitationsgeräusch *n*, Kavitationsrauschen *n*	bruits *mpl* de cavitation	cavitatiegeruis *n*
C 27	**cavitation nucleus**	Kavitationskeim *m*	noyau *m* de cavitation	cavitatiekern
C 28	**cavity**	Kavität *f*, Hohlraum *m*	cavité *f*	holte
C 29	**cavity resonance**	Gehäuseresonanz *f* <des Lautsprechers>, Hohlraumresonanz *f*	résonance *f* d'une cavité	holteresonantie
C 30	**celesta**	Celesta *n*, Stahlplattenklavier *n*	célesta *m*	celesta
C 31	**cembalo**	Cembalo *n*	cembalo *m*	cembalo
C 32	**cent** <musical interval>	Cent *n*	centième *m*	cent
C 33	**centibel**	Zentibel *n*	centibel *m*	centibel
C 34	**centi-octave**	Zentioktave *f*	centième *m* d'octave	centiem
C 35	**central (centre) frequency**, mean frequency	Mittenfrequenz *f*	fréquence *f* médiane (moyenne)	middenfrequentie
C 36	**centre-frequency error**	statischer Frequenzhub *m*	glissement *m* de fréquence médiane	verloop *n* van de middenfrequentie
C 37	**centre tuning**	Bandmittenabstimmung *f*	accord *m* de centre de bande	afstemming op het midden van de band
C 38	**ceramic microphone**	Kristallmikrofon *n*	microphone *m* à cristal	kristalmicrofoon
C 39	**certified hearing aid audiologist, CHAA**	geprüfter Hörhilfen-Audiologe *m*	auricologue-audiologue *m* approuvé	erkende hoortoestellen-audioloog
C 40	**chain amplifier**, distributed amplifier	Kettenverstärker *m*	amplificateur *m* en chaîne	tussenschakelversterker
C 41	**chamber music**	Kammermusik *f*	musique *f* de chambre	kamermuziek
C 42	**chancel**	Kanzelle *f* <Orgel>	laye *f*	windlade <orgel>
C 43	**channel**	Kanal *m*, Übertragungskanal *m*	canal *m*	kanaal *n*
C 44	**characteristic acoustical impedance**	Schallkennimpedanz *f*	impédance *f* acoustique caractéristique	karakteristieke akoestische impedantie
C 45	**characteristic band**	charakteristisches Frequenzband *n*	bande *f* caractéristique	karakteristieke band
C 46	**characteristic impedance**	Kennimpedanz *f*, Feldimpedanz *f*	impédance *f* caractéristique (de champ)	karakteristieke impedantie
C 47	**characteristic noise parameters**	charakteristische Rauschkenngrößen *fpl*	paramètres *mpl* de bruits parasitaires	ruisgetallen *npl*
C 48	**chattering**	Klappern *n*	cliquetis *m*, clapotis *m*	geratel *n*
C 49	**Chebishev shape**	Tschebyscheff-Charakteristik *f*	modèle *m* de Tchébycheff	Tsjebisjef-karakter *n*
C 50	**chest voice**	Bruststimme *f*	voix *f* de poitrine	borststem
C 51	**chime**	Glockenläuten *n*	carillon *m*, son *m* de cloches	engels carillon *n*
C 52	**chinese crash cimbals**	chinesische Becken *npl*	cimbales *fpl* chinoises	chinese bekkens *npl*
	chinese pavilion	*s.* B 105		
C 53	**chinese wood block**	Holzblocktrommel *f*	tambour *m* en bois chinois	houtbloktrommel
C 54	**chinking**	Klimpern *n*	cliquetis *m*	gerinkel *n*
C 55	**chirping**	Zwitschern *n*	gazouillis *m*	gepiep *n*
C 56	**Chladni's acoustic figures**	Chladnische Klangfiguren *fpl*	figures *fpl* de Chladni	figuren *npl* van Chladni
C 56a	**choir, chorus**	Chor *m*	chœur *m*, chorus *m*	koor *n*

	English	Deutsch	Français	Nederlands
C 57	choral	Choral *m*	choral *m*	koraal *n*
C 58	choral singing	Chorgesang *m*	chant *m* choral	koorzang
C 59	chord	Akkord *m*	accord *m*	akkoord *n*
	chorus	s. C 56a		
C 60	Christmas-tree pattern	Christbaumbild *n* <als Reflektogramm auf einer Schallplatte>	image *f* de bande <réflectogramme sur un disque de phonographe>	kerstboompatroon *n*
	Christmas-tree pattern wax picture	s. B 189		
C 61	chromatic	chromatisch	chromatique	gealtereerd akkoord *n*
	chromatic chord	s. A 190		
C 62	chromatic interval	chromatisches Intervall *n*	intervalle *m* chromatique	chromatisch interval *n*
C 63	chromatic scale	chromatische Tonleiter *f*	gamme *f* chromatique	chromatische toonladder
C 64	cimbalon, dulcimer	Hackbrett *n*	tympanon *m*, cymbale *f*	hakkebord *n*
C 65	circuit noise	Rauschen *n* <einer Schaltung>	bruit *m* de fond	elektronische ruis
C 66	circuit Q	Kreisgüte *f*	facteur *m* Q	kringkwaliteit
C 67	circular frequency, angular frequency, pulsatance	Kreisfrequenz *f*, Winkelfrequenz *f*	pulsation *f*, vitesse (fréquence) *f* angulaire	hoekfrequentie
C 68	circularly polarized sound wave	kreisförmig polarisierte Schallwelle *f*	onde *f* polarisée circulairement	geluidgolf met cirkelvormige polarisatie
C 69	circular radiation	Rundstrahlung *f*	rayonnement *m* circulaire	rondstraling
C 70	circumaural earphone	Kopfhörer *m* mit ohrumschließendem (circumauralem) Kissen	écouteur *m* à coussins enveloppants	circumaurale telefoon
C 71	circumferential oscillations	Umfangsschwingungen *fpl*	oscillations *fpl* circonférentielles	omtrekstrillingen *pl*
C 72	cittern	Kithara *f*	cithare *f*	citer
C 73	clang deafness	Klangtaubheit *f*	surdité *f* aux sons	klankdoofheid
C 74	clapper	Klöppel *m* <der Glocke>, Klapper *f*	battant *m* <de cloche>	klepel
C 75	clarinet	Klarinette *f*	clarinette *f*	klarinet
C 76	clarion	Oktavtrompete *f*	clairon *m*	klaroen
C 77	clarity <of tone>	Transparenz *f*, Durchsichtigkeit *f*	transparence *f* <d'un son>	klaarheid
C 78	clatter	rattern	cliqueter, trépider	klepperen
C 79	clavichord	Klavichord *n*	clavichorde *m*	clavichord *n*
C 80	clearness	Deutlichkeit *f*	clarté *f*	helderheid
C 81	click	knacken	craquer	klikken
C 82	clicks	Knackgeräusche *npl*	craquements *mpl*	geklik *n*
C 83	clipper	Schwellwertbegrenzer *m*	limiteur *m* à seuil	begrenzer
C 84	clipping	Abschneiden *n*, Beschneiden *n* <Amplitude>	limitation *f* d'amplitude	begrenzing
C 85	clipping of frequency range	Beschneiden *n* des Frequenzbereichs	limitation *f* de la plage de fréquences	afsnijden van het frequentiegebied
	CLL	s. C 113		
C 86	clockwise polarized <wave>	rechtsdrehend polarisiert	polarisé dextrogyre (à droite)	rechtsomdraaiend gepolariseerd
C 87	close	Kadenz *f*	cadence *f*	toonsluiting
C 88	close-end impedance	Blockimpedanz *f*, Kurzschlußimpedanz *f*	impédance *f* de court-circuit	kortsluitimpedantie
C 89	close-range effect	Nahwirkung *f*	effet *m* de proximité	nabijheidseffect *n*
C 90	close-range fading zone	Nahschwundzone *f*	zone *f* de fading proche	nabije overgangsgebied *n*
C 91	close-talking microphone	Nahbesprechungsmikrofon *n*	microphone *m* de proximité	handmicrofoon
C 92	close-talking response (sensitivity)	Übertragungsfaktor *m* für Nahbesprechung	facteur *m* de transmission proche	karakteristiek bij dichtbijbespreking
C 93	cloud	verdecken, abdecken <Töne>	couvrir <un son>	bedekken
C 94	clucking	Glucksen *n*	gloussement *m*	geklok *n*
C 95	coarse tuning	Grobstimmung *f*	accord *m* grossier	grof-afstemming
C 96	coated magnetic tape	Zweischichtenmagnetband *n*	ruban *m* magnétique à deux couches	beklede geluidband
C 97	coaxial loudspeaker	Koaxiallautsprecher *m*	haut-parleur *m* coaxial	coaxiale luidspreker
C 98	coaxial pad	koaxiales Dämpfungsglied *n*	atténuateur *m* coaxial	coaxiale verzwakker
C 99	cochlea	Schneckengang *m* <Ohr>	limaçon *m*	slakkenhuis *n*
	coefficient of absorption	s. A 10		
	coefficient of coupling	s. C 210		
C 100/1	coherer-type acoustic shock reducer	Frittersicherung *f*	réducteur *m* de la frittage	akoestische-schokdemper
C 102	cohering of the granules	Zusammenbacken *n* der Kohlekörner <im Mikrofon>	adhérence *f* des grains de charbon	aaneenkleven van de koolkorrels
C 103	coil-driven loudspeaker	dynamischer Lautsprecher *m*	haut-parleur *m* dynamique	dynamische luidspreker
C 104	coiled loudspeaker horn, twisted loudspeaker horn	gewundenes Lautsprecherhorn *n*	pavillon *m* contourné de haut-parleur	gedraaide luidsprekerhoorn
C 105/6	coincidence amplifier	Koinzidenzverstärker *m*	amplificateur *m* à coïncidence	coïncidentieversterker
C 107	coloration	Klangfärbung *f*	coloration *f*	kleuring
C 108	column loudspeaker	Säulenlautsprecher *m*	colonne *f* de haut-parleurs	kolomluidspreker
C 109	column speaker	Tonsäule *f*	colonne (console) *f* de haut-parleurs	luidsprekerkolom
C 110	combination microphone	Kombinationsmikrofon *n*	microphone *m* combiné	microfooncombinatie
C 111	combination tone	Kombinationston *m*	combinaison *f* de sons	combinatietoon
C 112	combination tone distortion	Intermodulationsverzerrung *f* <Funk>	distorsion *f* par intermodulation	combinatietoonvervorming
C 113	comfortable loudness level, CLL	angenehme Lautstärke *f*	niveau *m* sonore agréable	aangenaam geluidniveau *n*
C 114	common-mode rejection	Gleichtaktunterdrückung *f*	suppression *f* de phase commune	onderdrukking van gemeenschappelijk signaal
C 115	common-mode rejection quotient	Diskriminationsfaktor *m*	coefficient *m* de discrimination	verzwakking van gemeenschappelijk signaal
C 116	common reference	gemeinsamer Bezugspunkt *m*	référence *f* commune	gemeenschappelijke referentie

ID	English	German	French	Dutch
C 117	companding	Dynamikregelung f	réglage m d'expansion	dynamiekregeling
C 118	compandor	Kompander m	compendeur m	dynamiekregelaar
C 119	compensated amplifier	Kompensationsverstärker m	amplificateur m compensé	gecompenseerde versterker
C 120	compensated squelch	kompensierte Geräusch-sperre f	suppresseur m de bruit compensé	gecompenseerde ruisonder-drukking
C 121	compensated volume control	gehörrichtige Lautstärke-regelung f	contrôle m de volume compensé	gecompenseerde volumere-geling
C 122	compensation of filter attenuation	Ausgleich m der Filter-dämpfung	compensation f d'atténua-tion de filtre, compensa-tion de pertes de filtrage	compensatie van filterver-zwakking
C 123	compensation of static pressure increase	Druckstauentzerrung f <Mikrofon>	compensation f de crois-sance de pression statique	statische drukcompensatie
C 124	complete cycle	volle Periode f	période f pleine, cycle m complet	gehele periode
C 125	completely diffuse sound	vollständig diffuser Schall m	son m totalement diffus	volledig diffuus geluid n
C 126	completely matched power gain	maximal verfügbare Leistungsverstärkung f	puissance f d'amplification maximum disponible	optimaal aangepaste vermo-gensversterking
C 127	complete modulation	Vollaussteuerung f	modulation f totale	volledige modulatie
C 128	complex admittance	komplexe Admittanz f	admittance f complexe	complexe admittantie
C 129	complex excitation	komplexe Erregung f	excitation f complexe	complexe aanstoting
C 130	complex quantity	komplexe Größe f	quantité f complexe	complexe grootheid
C 131	complex response	komplexe Belastung f <einer Quelle>	charge f complexe	complexe responsie
C 132	complex sound	komplexer Schall m	son m complexe	samengesteld geluid n
C 133	complex system param-eter	komplexe Impedanz f <eines linearen Systems>	impédance f complexe	complexe impedantie
C 134	complex tone	Tongemisch n	mélange m de sons	samengestelde toon
C 135	compliance	Nachgiebigkeit f, reziproke Steifigkeit f	compliance f	compliantie
C 136	component frequency	Teilfrequenz f	fréquence f composante	frequentiecomponent
C 137	composite attenuation	Betriebsdämpfung f	atténuation f complexe	samengestelde verzwakking
C 138	composite loudspeaker	Lautsprecherkombination f	ensemble m de haut-parleurs	luidsprekercombinatie
C 139	composite noise ex-posure index	kombinierter Lärmexpo-nierungsindex m	index m de bruits composé	samengestelde lawaaibe-lastingsindex
C 140	compressed speech	gepreßte Sprache f	parole f compressée	gecomprimeerde spraak
C 141	compressional wave	Kompressionswelle f	onde f de compression	drukgolf
C 142	compression amplifier	Verstärker m mit Dynamik-pressung	amplificateur m à compres-sion	versterker met dynamiek-compressie
C 143	compression of band	Bandpressung f, Bandein-engung f	compression f de bande	bandcompressie
C 144	compression of volume, constriction of volume	Dynamikeinengung f	compression f de volume	volumecompressie
C 145	compressor	Dynamikpresser m	compresseur m de volume	compressor
C 146	concentrate	bündeln	concentrer	bundelen
C 147	concert flute	große Flöte f	grande flûte f	grote fluit
C 148	concert grand	Konzertflügel m	piano m à queue	concertvleugel
C 149	concert horn, French horn	Waldhorn n	cor m de chasse, cor d'har-monie	waldhoorn
C 150	concertina	Konzertina f	concertina f	concertina
C 151	concert pitch	Stimmton m, Orchester-abstimmton m	ton-étalon m	stemtoon
C 152	concord	konsonanter Akkord m	accord m consonant	welluidend akkoord n
C 153	concrete music	konkrete Musik f	musique f concrète	concrete muziek
	condenser loudspeaker	s. C 8		
	condenser microphone	s. C 9		
C 154	conditioned reflex, CR	bedingter Reflex m	réflexe m conditionné (conditionnel)	geconditioneerde reflex
C 155	conditioned stimulus, CS	bedingter Reiz m	excitation f conditionnelle	geconditioneerde stimulus
C 156	conductance	Konduktanz f	conductance f	conductantie
C 157	conductive deafness	Leitungstaubheit f	surdité f de conduction	geleidingsdoofheid
C 158	conductive ripple pick-up	galvanische Brummschleife f	couplage m galvanique de ronflement	geleidingsbrom
C 158a	cone diaphragm	Konusmembran f	membrane f conique	conusmembraan
C 159	cone loudspeaker	Konuslautsprecher m, Trichterlautsprecher m	cône m de haut-parleur	conusluidspreker
C 160	conference repeater	Konferenzverstärker m	amplificateur m d'interphone	interfoonversterker
C 161	conical horn	Kegeltrichter m	pavillon m conique	conische hoorn
C 162	conjugate impedance	konjugiert komplexe Impedanz f	impédance f conjuguée complexe	toegevoegd complexe impe-dantie
C 163	conservative flux	quellenfreier Fluß m	flux m exempt de source	conservatieve flux
C 164	consonant articulation	Konsonantenverständlich-keit f	compréhension f de conson-nes	verstaanbaarheid van mede-klinkers
C 165	constancy of pitch	Tonhöhenkonstanz f	constance f de hauteur de son	constantheid van de toon-hoogte
C 166	constante-amplitude recording	Schallaufzeichnung f mit konstanter Amplitude	enregistrement m de son à hauteur constante	geluidregistratie met con-stante amplitude
C 167	constant-percentage bandwidth	konstante relative Band-breite f	largeur f de bande propor-tionnellement constante	relatief constante band-breedte
C 168	constant-velocity recording	Schallaufzeichnung f mit konstanter Schnelle	enregistrement m à vélocité constante	geluidregistratie met constante snelheid
	constriction of volume	s. C 144		
C 169	contact microphone	Kontaktmikrofon n	microphone m à contact	contactmicrofoon
C 170	contact-modulated amplifier	Verstärker m mit Eingangs-unterbrechung	amplificateur m à modula-tion de contact	versterker met contactmo-dulatie
C 171	continuous <wave>	gleichförmig, gleichleitend, ungedämpft	uniform, constant	continu
C 172	continuous frequency audiometer	Frequenzstufenaudiometer n	audiomètre m à fréquences graduées	audiometer met continue frequentie
C 173	continuous oscillation	ungedämpfte (gleich-förmige) Schwingung f	oscillation f continue (non amortie)	ongedempte trilling
C 174	continuous sound, steady sound, sustained tone	Dauerton m, ausgehaltener Ton m	son m continu, ton m sou-tenu	continu geluid n
C 175	continuous sound level	Dauerlärmpegel m, Dauer-schallpegel m	niveau m de bruit continu	continu geluidpeil n

C 176	continuous spectrum	kontinuierliches Spektrum n	spectre m continu	continu spectrum n
C 177	continuous spectrum noise	Geräusch n mit kontinuierlichem Spektrum	spectre m de bruit continu	ruis met continu spectrum
C 178	continuous system	Kontinuum n	système m continu	continu systeem n
C 179	contrabass, double-bass	Baßgeige f, Kontrabaß m	contrebasse f à cordes	contrabas
C 180	contrabass clarinet, double-bass clarinet	Kontrabaßklarinette f	clarinette-contrebasse f	contrabasklarinet
C 181	contra bassoon, double bassoon	Kontrafagott n	contrebasson m	contrafagot
C 182	contrabass trombone, double-bass trombone	Kontrabaßposaune f	trombone-contrebasse m, contrebasse f à vent	contrabastrombone
C 183	contralateral routing of signals <in hearing aids> CROS	Signalumlenkung f zum gesunden Ohr <in Hör­hilfen>	report m du signal sur l'oreille fonctionnelle <dans une aide acoustique>	omleiding van signalen naar het andere oor
C 184	contralto	Kontraalt m	contralto m, contralte m	alt
C 185	contrapuntal	kontrapunktisch, polyphon	contrapuntique	contrapuntisch
C 186	contrast amplification	Verstärkung f unter Dynamikdehnung	amplification f de contraste	dynamiekversterking
C 187	contrast control	Dynamikregelung f	réglage m de contraste	dynamiekregeling
C 188	contrast expansion, expansion	Dynamikdehnung f	expansion f [de contraste]	dynamiekexpansie
C 189	control setting	Reglerstellung f	position f (positionnement m) de contrôle	instelling van de regelknoppen
C 190	control track	Kontrollspur f <auf Magnet­band oder Film>	trace f de contrôle	controlespoor n
C 191	convergence zone	Konvergenzzone f	zone f de convergence	convergentiezone
C 192	converging wave	konvergierende Welle f	onde f convergente	convergerende golf
C 193	conversion chart for vibration parameters	Schwingungskarte f	carte f de conversion des paramètres d'oscillations	conversiegrafieken pl voor trillingspatronen
C 194	conversion deafness	hysterische Taubheit f	surdité f hystérique (de conversion)	hysterische doofheid
C 195	cord	Saite f	corde f	snoer n
C 196	cordless ultrasonic control	Ultraschallfernbedienung f <TV>	télécommande f par ultrasons	draadloze ultrasonore bediening
C 197	corner horn	Ecklautsprecher m	haut-parleur m de coin, haut-parleur d'encoignure	hoekluidspreker
C 198	cornet	Kornett n	cornet m	kornet
C 199	corona loudspeaker	Koronalautsprecher m	haut-parleur m ionique	ionofoonluidspreker
C 200	cosine wave	Kosinuswelle f	onde f sinusoidale	sinusvormige golf
C 201	Coulomb damping	Coulombsche Dämpfung f	amortissement m de Coulomb	Coulomb-demping
C 202	counterclockwise polarized wave, left-handed polarized wave	linksdrehend (elliptisch) polarisierte Welle f	onde f lévogyre (polarisée à gauche)	linksdraaiend gepolariseerde golf
C 203	counter-coupling, inverse feedback, negative reaction	negative Rückkopplung f, Gegenkopplung f	réaction f négative, contre-réaction f	tegenkoppeling
C 204	counter part	Gegenstimme f <Musik>	contre-chant m	tegenstem
C 205	counter point	Kontrapunkt m	contrepoint m	contrapunt n
C 206	coupled-circuit effect	Doppelhöckereffekt m	effet m de circuits surcouplés	effect n van gekoppelde kringen
C 207	coupled impedance	Kernwiderstand m	impédance f couplée	gekoppelde impedantie
C 208	coupled modes	gekoppelte Moden mpl, gekoppelte Eigenschwingungen fpl	modes mpl couplés	gekoppelde trillingswijzen pl
C 209	coupler couplet	Kuppler m, Koppler m s. D 212	coupleur m	koppelstuk n
C 210	coupling coefficient (factor), coefficient of coupling	Koppelfaktor m, Koppelkoeffizient m	facteur (coefficient) m de couplage	koppelfactor
C 211	coupling loss	Anpassungsverlust m, Belastungsverlust m <eines Schallaufnehmers>	pertes fpl d'adaptation	aanpassingsverlies n
C 212	coupling transformer	Kopplungsübertrager m	transformateur m de couplage	koppeltransformator
C 213	covibration	Resonanzschwingung f, Koinzidenzschwingung f	oscillations fpl de résonance	meetrilling
	CR	s. C 154		
C 214	crack	knacken	craquer	kraken
C 215	crackle	knistern	crépiter	geknetter n
C 216	crackling noise	Knistergeräusch n	bruit m de crépitement	gekraak n
C 217	crash	Knall m	claquement m	geraas n
C 218	creak	knarren	craquer, grincer	knarsen
C 219	crest factor	Scheitelfaktor m	facteur m de crête	piekfactor
C 220	crest of oscillation	Schwingungsmaximum n, Schwingungsscheitel m	crête f d'oscillation	topwaarde van de trilling
C 221	criss-cross system	Hörhilfe f mit Verstärker-Mikrofon-Kombination in gekreuzter Anordnung	aide f acoustique à branchement croisé	hoortoestel n met gekruiste verbindingen
C 222	critical damping	kritische Dämpfung f	amortissement m critique	kritische demping
C 223	critical frequency	Resonanzfrequenz f <bei Schwingungsanregung>	fréquence f de résonance	kritische frequentie
C 224	critical speed	kritische Drehzahl f (Umdrehungsfrequenz f)	vitesse f critique	kritische snelheid
C 225	croaking	Krächzen n	croassement m	gekras n
C 226	crock	Stimmbogen m	corps m de rechange	stembeugel
C 227	cromorna, crumhorn, krummhorn	Krummhorn n <Musik>	cromorna m <registre d'orgue>	kromhoorn
	CROS	s. C 183		
C 228	cross drift	Querdrift f	dérive f de couplage	uit het spoor lopen
C 229	cross fading	Überblenden n	enchainement m	overvloeien
C 230	cross flute, German flute, transverse flute	Querflöte f	flûte f traversière	dwarsfluit
C 231	cross modulation	Kreuzmodulation f	intermodulation f	kruismodulatie

C 232	cross-over frequency	Überlappungsfrequenz *f*	fréquences *fpl* de jonction	overgangsfrequentie
C 233	cross-over point	Überlappungspunkt *m*	point *m* focal (de jonction)	overgangspunt *n*
C 234	cross-over range	kritische Distanz *f*	distance *f* critique (de convergence)	kritische afstand
C 235	cross-over spiral	Überspringauslenkung *f* <auf Schallplatte>	sautage *m* de sillons	overgangsgroef
C 236	cross section of the beam	Strahlquerschnitt *m*	section *f* du faisceau	bundeldoorsnede
C 237	cross stroke	Querbalken *m* <Musik>	barre *f* de croches	verbindingsstreep van achtste noten
C 238	crosstalk	Übersprechen *n*	diaphonie *f*	overspraak
	crowd noise	*s.* A 296		
	crumhorn	*s.* C 227		
C 239	crystal diaphragm	Kristallmembran *f*	diaphragme *m* à cristal	kristalmembraan *n*
C 240	crystal loudspeaker, piezoelectric loudspeaker	piezoelektrischer Lautsprecher *m*	haut-parleur *m* à cristal	piëzo-elektrische luidspreker
C 241	crystal microphone, piezoelectric microphone	Kristallmikrofon *n*	microphone *m* à cristal	piëzo-elektrische microfoon
	CS	*s.* C 155		
C 242	cueing device	Regieeinrichtung *f* <Mischpult>	dispositions *fpl* de régie	regiehulpmiddel *n*
C 243	cue track	Regiespur *f*	plage *f* de régie	regiespoor *n*
	curtal	*s.* B 70		
C 244	curve of decay	Abklingkurve *f*	courbe *f* d'évanouissement	afnamecurve
C 245	cut-off frequency, limit frequency	Grenzfrequenz *f*	fréquence *f* limite (de coupure)	afsnijfrequentie
C 246	cut-off wavelength	kritische Wellenlänge *f*	longueur *f* d'onde de coupure	grensgolflengte
C 247	cut-over	Überschneidung *f*	recouvrement *m*	snijdersprong
C 248	cutting stylus	Schneidstift *m*, Schneidstichel *m*	stylet *m* graveur	snijbeitel, snijnaald
C 249	cycle	Zyklus *m*, Schwingungsperiode *f*, Periode *f*	cycle *m*, période *f*	periode
C 250	cylindrical wave	zylindrische Welle *f*	onde *f* cylindrique	cilindrische golf
C 251	cymbal	Zimbel *f*	cymbale *f*	bekken *n*, cimbaal
C 252	cymbals	Becken *npl* <Musik>	cymbales *fpl*	bekkens *npl*

D

D 1	dactyl speech	Fingersprache *f*, Taubstummensprache *f*	langage *m* des sourds-muets	gebarentaal
	DAF	*s.* D 50		
D 1a	damp	dämpfen	amortir	dempen
D 2	damped harmonic oscillation	gedämpfte harmonische Schwingung *f*	oscillation *f* harmonique amortie	gedempte harmonische trilling
D 3	damped impedance	Scheinwiderstand *m* eines gedämpften Systems	impédance *f* d'un système amorti	impedantie van een gedempt systeem
D 4	damped sinusoidal quantity	gedämpfte Sinusgröße *f*	grandeur *f* sinusoïdale amortie	gedempte sinusvormige grootheid
D 5	damped tuned circuit	gedämpft abgestimmter Kreis *m*	circuit *m* accordé amorti	gedempte afgestemde kring
D 6	damper, mute	Dämpfer *m* <des Musikinstruments>	sourdine *f*	demper
D 7	damper pedal, loud pedal	Forte-Pedal *n*	pédale *f* forte	forte-pedaal *n*, open pedaal *n*
D 8	damping, muting	Dämpfung *f*, Abschwächen *n*	amortissement *m*, assourdissement *m*	demping
D 9	damping factor	Dämpfungsfaktor *m*	facteur *m* d'amortissement	dempingsfactor
D 10	damping of air oscillation	Dämpfung *f* der Luftschwingung	amortissement *m* des oscillations de l'air	demping van luchttrillingen
D 11	damping ratio	Dämpfungsgrad *m*	degré *m* d'amortissement	dempingsfactor
D 12	damping torque	Dämpfungsmoment *n*	moment *m* amortisseur	dempingsmoment *n*
	dB	*s.* D 31		
D 13	DC noise voltage	Gleichfeldrauschspannung *f*	tension *f* de bruit en courant continu	gelijkspanning van de ruis
D 14	dead acoustics	trockene Akustik *f*	acoustique *f* sèche (sourde)	droge akoestiek
D 15	deaden	abdämpfen <Instrument>	mettre en sourdine, assourdir	dempen
D 16	deadness	Trockenheit *f*	sécheresse *f*, surdité *f*	droogheid
D 17	dead room	schalltoter Raum *m*	chambre *f* sourde	echoloze kamer
D 18	dead studio	trockenes (schalltotes) Studio *n*	studio *m* sec (sourd)	galmvrije studio
D 19	deaf	taub	sourd	doof
	deaf aid	*s.* H 38		
	deaf-and-dump	*s.* D 21		
D 20	deafen	ertauben	assourdir	verdoven
D 21	deaf-mute, deaf-and-dump	taubstumm	sourd-muet	doofstom
D 22	deaf-mutism	Stummheit *f* infolge Taubheit	mutisme *m* par suite de surdité	doofstomheid
D 23	deafness	Schwerhörigkeit *f*, Taubheit *f*	surdité *f*	doofheid
D 24	decay	abklingen, ausschwingen	diminuer, évanouir, décroître	uittrillen
D 25	decay	Abfall *m*	diminution *f*, évanouissement *m* <d'un son>, décroissement *m*	uittrilling
D 26	decay characteristic	Abklingcharakteristik *f*	caractéristique *f* d'amortissement	uittrillingsfactor
	decay coefficient	*s.* A 253		
D 27	decaying oscillation, dying oscillation	abklingende Schwingung *f*	oscillation *f* décroissante	gedempte trilling
D 28	decay of sound	Ausklingen *n* <des Schalles>	décroissement *m* du son	uitklinken

	English	German	French	Dutch
D 29	**decay rate**, rate of decay	Abklingrate f, Abklingdauer f	degré m d'évanouissement, degré de dégression	afname
D 30	**decay time**	Abklingzeit f, Abfallzeit f	temps m d'évanouissement	wegsterftijd
D 31	**decibel, dB**	Dezibel n, dB n	décibel m	decibel
D 32	**decouple**	entkoppeln	découpler	ontkoppelen
D 33	**decoupling**	Entkopplung f	decouplement m, découplage m	ontkoppeling
D 34	**decoupling network**	Entkoppler m, Entkopplungsschaltung f	réseau m de découplage	ontkoppelingsnetwerk n
D 35	**decrease of gain**	Verstärkungsabfall m	perte f d'amplification	versterkingsvermindering
D 36	**decreasing amplitude**	abnehmende Amplitude f	amplitude f décroissante	afnemende amplitude
D 37	**de-emphasizing filter**	Entzerrungsfilter n	filtre m compensateur	terugduwfilter n
D 38	**deep fade**	Tiefschwund m	perte f de graves	verzwakking van de allerlaagste tonen
D 39	**deep scattering layer**	Tiefenstreuschicht f	couche f de diffusion basse	diepe verstrooiende laag
D 40	**deflected beam**	abgelenkter Strahl m	rayon m dévié	afgebogen bundel
D 41	**deflecting baffle of loudspeaker**	Lautsprecherwand f	baffle m de haut-parleur	richtende klankkkaatser
D 42	**deflecting labyrinth of loudspeaker**, labyrinth of loudspeaker	Lautsprecherlabyrinth n	labyrinthe m de haut-parleur	akoestisch labyrint n in een luidsprekerkast
D 43	**deflection by wind**	Windablenkung f	déflecteur m de vent	afbuiging door de wind
	degenerative amplifier	s. N 17		
D 44	**degenerative feedback**, reverse feedback	Gegenkopplung f	contre-réaction f	tegenkoppeling
D 45	**degree of coupling**	Kopplungsgrad m	degré m de couplage	mate van koppeling
D 46	**degree of freedom**	Freiheitsgrad m	degré m de liberté	vrijheidsgraad
D 47	**degree of matching**	Anpassungsgrad m	degré m d'adaptation	mate van aanpassing
D 48	**degree of modulation**	Modulationsfaktor m, Aussteuerungsgrad m	degré m de modulation	modulatiediepte
D 49	**degree of uniformity**	Gleichförmigkeitsgrad m	degré m d'uniformité	mate van gelijkvormigheid
D 50	**delayed auditory feedback, DAF**	verzögert akustische Rückkopplung f	audition f différée en retour	vertraagde gehoorterugkoppeling
D 51	**delayed automatic gain control**	verzögerte automatische Verstärkungsregelung f	réglage m d'amplification retardé	vertraagde automatische sterkteregeling
D 52	**delayed feedback circuit**	verzögerte Rückkopplungsschaltung f	circuit m à réaction retardée	schakeling voor vertraagde terugkoppeling
D 53	**delay network**	Verzögerungsschaltung f	réseau m de délai	vertragingslijn
D 54	**demodulation**, detection	Entmodelung f, Demodulation f	démodulation f, détection f	demodulatie, detectie
D 55	**demodulation stage**, detector stage	Demodulatorstufe f	étage m démodulateur (de démodulation)	demodulatietrap
D 56	**demodulator circuit**	Demodulatorschaltung f	circuit m démodulateur	demodulatieschakeling
D 57	**densitometer**	Dichtemesser m	densitomètre m	densitometer
D 58	**descant**	Diskant m	discant m, soprano m	discant
D 59	**descending loudness level**	abnehmender Lautstärkepegel m	niveau m sonore décroissant	dalend geluidpeil n
D 60	**detection**, perception	Wahrnehmung f, Empfindung f	perception f, detection f	perceptie, waarneming
	detection	s. a. D 54		
D 61	**detection differential**, differential threshold	Wahrnehmbarkeitsschwelle f, Wahrnehmbarkeitsstufe f	seuil m de détection	detectiedrempel
D 62	**detectophone**	Lauschmikrofon n	microphone m de détection	afluistermicrofoon
D 63	**detector**	Demodulator m	démodulateur m	detector, demodulator
	detector stage	s. D 55		
D 64	**detunability**	Verstimmbarkeit f	désaccordabilité f	verstembaarheid
D 65	**detune**, mistune	verstimmen	désaccorder	verstemmen
D 66	**detuning**, mistuning	Verstimmung f	désaccord m	verstemming
D 67	**detuning measurement**	Verstimmungsmessung f	mesure f de désaccord	meting van de verstemming
D 68	**deviation**	Abweichung f	déviation f	afwijking
D 69	**dextropropagating wave**	rechtslaufende Welle f	onde f dextrogyre	rechtsom lopende golf
D 70	**diaphragm**	Diaphragma n, Membran f	diaphragme m, membrane f	membraan n
D 71	**diaphragm cap**	Membrandeckel m, Membrankappe f	cape f de diaphragme	membraankapje n
D 72	**diaphragm displacement**	Membranauslenkung f	déviation f du diaphragme	uitwijking van het membraan
D 73	**diaphragmless microphone**	membranloses Mikrofon n	microphone m sans membrane	membraanloze microfoon
D 74	**diaphragm oscillation**	Membranschwingung f	oscillation f de diaphragme	membraantrilling
D 75	**diaphragm stiffness**	Membransteifigkeit f	raideur f de diaphragme	stijfheid van het membraan
D 76	**diatonic**	diatonisch	diatonique	diatonisch
D 77	**diatonic interval**	diatonisches Intervall n	intervalle m diatonique	diatonisch interval n
D 78	**diatonic system**	Diatonik f	système m diatonique	diatonisch systeem n
D 79	**die away**	ausklingen, ausschwingen	décroître, s'évanouir	wegsterven
D 80	**die out**	verklingen	s'évanouir	verklinken
D 81	**difference frequency**	Differenzfrequenz f	fréquence f différentielle	verschilfrequentie
D 82	**difference limen, DL**	Wahrnehmbarkeitsstufe f	limite f de perception	trede
D 83	**difference limen for loudness**	Wahrnehmbarkeitsstufe f für Lautheit	limite f de différentiation d'intensité d'un son	intensiteitstrede
D 84	**difference limen for pitch**	Wahrnehmbarkeitsstufe f für Tonhöhe	limite f de différentiation de hauteur de ton	toonhoogtetrede
D 85	**difference tone**	Differenzton m	ton m différentiel	verschiltoon
D 86	**differential microphone**	Differentialmikrofon n	microphone m différentiel	differentiaalmicrofoon
D 87	**differential phase**	differentielle Phase f	phase f différentielle	onderling faseverschil n
	differential threshold	s. D 61		
D 88	**differentiating network**	Differentiator m	réseau m différentiateur	differentiëerschakeling
D 89	**diffract**	beugen <Welle>	diffracter	buigen
D 90	**diffracted wave**	Beugungswelle f	onde f diffractée	afgebogen golf
D 91	**diffraction**	Beugung f	diffraction f	buiging
D 92	**diffraction beam**	Beugungsstrahl m	rayon m diffracté	bundel afgebogen golven
D 93	**diffraction fringes**	Beugungsfransen fpl	franges fpl de diffraction	buigingspatroon n
D 94	**diffraction losses**	Beugungsverluste mpl	pertes fpl de diffraction	buigingsverlies n
D 95	**diffraction region**	Beugungszone f	zone f de diffraction	buigingsgebied n

D 96	diffuse field	diffuses Schallfeld n	champ m diffus	diffuus veld n
D 97	diffuse-field distance	Hallradius m	rayon m du champ de diffusion	galmstraal
D 98	diffuser	Streuelement n, Streuschirm m	diffuseur m	verstrooier
D 99	diffuse reflection	diffuse Reflexion f	réflexion f diffuse	diffuse terugkaatsing
D 100	diffuse sound	diffuser Schall m	son m diffus	diffuus geluid n
D 101	diffusing surface	diffus streuende Oberfläche f	surface f diffusante	diffuus verstrooiend oppervlak n
D 102	diffusivity	Diffusität f	diffusibilité f, diffusivité f	diffusiteit
D 103	dilatational wave	Dehnwelle f	onde f de dilatation	dilatatiegolf
D 104	diminished third	verminderte Terz f	tierce f mineure	verkleinde terts
D 105	ding	klingeln	sonner	bellen
D 106	dip	Schwingungsloch n	trou m d'oscillations, dip m	dip
D 107	direct	richten	diriger, orienter	richten
D 108	direct-coupled	galvanisch gekoppelt	à couplage galvanique	galvanisch verbonden
D 109	directed beam	gerichteter Strahl m, Richtstrahl m	rayon m dirigé	gerichte bundel
D 110	directional characteristic	Richtungsaufnahmecharakteristik f	caractéristique f directionnelle	richtingsgevoeligheidskarakteristiek
D 111	directional couple	Direktionsmoment n	couple m directionnel	richtkoppel n
D 112	directional distribution of sound	Schallrichtungsverteilung f	répartition f directionnelle du son	ruimtelijke verdeling van het geluid
D 113	directional gain, directivity index	Bündelungsmaß n	gain m directionnel, gain (index m) de directivité	bundelingsfactor
D 114	directional loudspeaker	Lautsprecher m mit Richtwirkung, Richtlautsprecher m	haut-parleur m directionnel	gerichte luidspreker
D 115	directional microphone	Richtmikrofon n	microphone m directionnel	gerichte microfoon
D 116	directional perception	Richtungswahrnehmung f	perception f directionnelle	richtinghoren
D 117	directional reception	Richtempfang m	réception f directive	gerichte ontvangst
D 118	directional response pattern	Richtcharakteristik f eines Wandlers	diagramme m de réponse directionnelle	richtingsdiagram n
	directional response (sensitivity) pattern	s. D 123		
D 119	directional transmitter	Richtstrahler m	émetteur m directif	gerichte bron
D 120	direction discrimination	Richtungsunterscheidung f	discrimination f de direction	richtingsonderscheid n
D 121	directive screen	Perlwand f	écran m à perles	scherm n met gerichte terugkaatsing
D 122	directive wave	gerichtete Welle f	onde f dirigée	gerichte golf
D 123	directivity characteristic, directional response (sensitivity) pattern, beam pattern	Richtcharakteristik f	caractéristique f de directivité	richtingsdiagram n
D 124	directivity factor directivity index	Bündelungsgrad m s. D 113	facteur m de directivité	bundelingsfactor
D 125	directivity pattern polar response	Richtungscharakteristik f	diagramme m polaire de directivité	polair richtingsdiagram n
D 126	direct recording	unmittelbare (direkte) Schallaufzeichnung f	enregistrement m direct	directe registratie
D 127	discord, dissonance	Dissonanz f	dissonance f	dissonant
	discrete frequency audiometer	s. P 199		
	discrete-parameter system	s. L 126		
	discrete word intelligibility	s. A 238		
D 128	discriminate against harmonics	Harmonische ausfiltern (aussieben)	filtrer les harmoniques	discrimineren tegen harmonischen
D 129	discrimination loss	Unterscheidungsverlust m <für Spracherkennung>	perte f de discrimination	discriminatieverlies n
	discriminator	s. F 83		
D 130	discriminator score for speech	Unterscheidungsgrad m für Sprache	degré m de discrimination de la parole	mate van spraakverstaanbaarheid
D 131	disharmonious, inharmonious	disharmonisch, unharmonisch	disharmonieux, discordant	disharmonisch
D 132	disharmony	Disharmonie f	disharmonie f	disharmonie
D 133	disk	Schallplatte f	disque m	plaat
D 134	dispersed magnetic powder tape	Masseband n, Massemagnetband n	bande f magnétique massive	door-en-door magnetische band
D 135	dispersed magnetic powder tape	Vollpulvermagnetband n, Pulvermagnetband n	bande f magnétique à poudre dispersée	geluidband met magnetisch poeder
D 136	dispersion	Dispersion f	dispersion f	dispersie
D 137	displacement	Ausschlag m	déviation f, déplacement m	verplaatsing
D 138	displacement amplitude	Ausschlagamplitude f	amplitude f de déviation (déflexion)	verplaatsingsamplitude
D 139	displacement transducer	Ausschlagaufnehmer m	transducteur m de déflexion	verplaatsingstransducent
D 140	dissipation	Dissipation f	dissipation f	dissipatie
D 141	dissipation coefficient	Dissipationsgrad m	coefficient m de dissipation	dissipatiefactor
	dissonance	s. D 127		
D 142	dissymmetrical transducer	asymmetrischer Wandler m	transducteur m dissymétrique	asymmetrische transducent
D 143	distinct	deutlich	distinct	duidelijk
D 144	distinctive	gut erkennbar, charakteristisch	distinctif	kenmerkend
D 145	distinctness	Deutlichkeit f, Erkennbarkeit f	distinctibilité f	duidelijkheid
D 146	distort	verzerren	distordre	vervormen
D 147	distorted waveform	verzerrte Wellenform f	onde f distordue	vervormde golf
D 148	distortion	Verzerrung f	distorsion f	distorsie
D 149	distortion analyser	Klirranalysator m	analyseur m de distorsion	vervormingsanalysator
D 150	distortion correction	Entzerrung f	compensation f (correction f) de distorsion	vervormingscorrectie
D 151	distortion factor, harmonic distortion	Klirrfaktor m	facteur m de distorsion	vervormingsfactor

D 152	distortion factor of voice channel	Gesprächsklirrfaktor m, Sprechkanalklirrfaktor m	facteur m de distorsion d'un canal de transmission	vervormingsfactor in het spraakkanaal
D 153	distortion-free, distortionless	verzerrungsfrei, unverzerrt, klirrfrei	sans distorsion, non distordu	vervormingsvrij
D 154	distortionless microphone	verzerrungsfreies Mikrofon n	microphone m non distordant	vervormingsvrije microfoon
D 155	distortion of sound	Lautverzerrung f	distorsion f de son	geluidvervorming
D 156	distortion of sound field	Schallfeldverzerrung f	distorsion f de champ acoustique	vervorming van het geluidveld
	distributed amplifier	s. C 40		
D 157	distributed system	System n mit unendlich vielen Freiheitsgraden	système m avec un nombre infini de degrés de liberté	systeem n met eindeloos veel vrijheidsgraden
	disturbing field	s. I 101		
	disturbing wave	s. I 106		
D 158	divergence loss, spreading loss	Divergenz-Dämpfungsmaß n, Verlust m durch divergierende Schallausbreitung	pertes fpl par divergence	spreidingsverlies n
D 159	divide-by-two circuit	Frequenzhalbierschaltung f <für Lautsprecher>	circuit m d'aiguillage de fréquences	frequentiedelingsschakeling
D 160	divider chain	Teilerkette f	chaîne f diviseuse	delerketen
D 161	dividing network	Frequenzbandteilerschaltung f <für Lautsprecher>, Aufteilungsschaltung f	réseau m répartiteur de fréquences	delernetwerk n
	DL	s. D 82		
D 162	domestic listening conditions	Hörbedingungen fpl in Wohnräumen	conditions fpl auditives domestiques	huiskamercondities pl
D 163	dominant chord	Dominantenakkord m	accord m de dominante	dominant akkoord n
D 164	dominant frequency	Hauptfrequenz f	fréquence f dominante	dominante frequentie
D 165	Doppler effect	Doppler-Effekt m	effet m de Doppler	dopplereffect n
D 166	Doppler shift	Doppler-Verschiebung f, Frequenzverschiebung f infolge des Doppler-Effekts	glissement m de fréquence par effet Doppler	dopplerverschuiving
D 167	double amplitude	Spitze-Spitze-Wert m der Sinusschwingung	amplitude f pointe-pointe	top-topwaarde van de trilling
D 168	double bar	doppelter Taktstrich m	double barre f	dubbele maatstreep
	double-bass	s. C 179		
	double-bass clarinet	s. C 180		
	double bassoon	s. C 181		
	double-bass trombone	s. C 182		
D 169	double-button carbon microphone	Differentialkohlemikrofon n	microphone m différentiel (symétrique) à grenaille	differentiaalkoolmicrofoon
D 170	double diapason	Doppelprinzipal m	diapason m double	dubbelprestant
D 171	double-edged variable width sound track	Doppelzackenschrift f, Doppelzackenschallaufzeichnung f	gravure f par élongation variable double	dubbelsporige geluidfilm met variabele spoorbreedte
D 172	double flat	Doppel-b n	double bémol m	dubbelmol
D 173	double fugue	Doppelfuge f	double fugue f	dubbele fuga
D 174	double-hump curve	zweihöckrige Kurve f	courbe f en double bosse	grafiek met twee toppen
D 175	double-hump effect	Doppelhöckereffekt m	effet m de double bosse	effect n van een curve met twee toppen
D 176	double-peak resonance curve	zweispitzige Resonanzkurve f	courbe f de résonance à double pic	resonantiecurve met twee toppen
D 177	double regeneration	zweifache Rückkopplung f	double réaction f	dubbele terugkoppeling
D 178	doubler stage	Verdopplerstufe f	étage m doubleur	verdubbelingstrap
D 179	double sharp	Doppelkreuz n	double dièze m	dubbel kruis n
D 180	double side band	Zweiseitenband n	doubles bandes fpl latérales	dubbele zijband
D 181	double stopping	Doppelgriff m <Musik>	double position f	dubbelgreep
D 182	double tonguing	Doppelzungenschlag m	doubel coup m de langue	dubbele tongslag
D 183	double touch	Doppelpedal n	pédale f double	combinatiepedaal n
D 184	double vibration	Doppelschwingung f	vibration f double	dubbele trilling
D 185	double whole note	Brevis f	brève f	dubbelhele noot
D 186	double-width push-pull sound print	doppelt breite Gegentakt-Tonkopie f	copie f sonore push-pull double format	dubbelbreed balansgeluidspoor n
D 187	doubling of frequency	Frequenzverdopplung f	doublage m de fréquence	frequentieverdubbeling
	DR	s. D 222		
D 188	drawknob	Registerknopf m <Orgel>, Zugknopf m	boutons mpl de jeux	registerknop
D 189	drawl	gedehnt und langsam sprechen	parler lentement en trainant les mots	temen
D 190	drawn string	gespannte Saite f	corde f tendue	gespannen snaar
D 191	driving-point impedance	Antriebspunktimpedanz f	impédance f au point de réglage	ingangsimpedantie
D 192	drop-out	Ausfall m <des Signals>	absence f de signal	gat n
D 193	drum	Trommel f	tambour m	trom
D 194	drumhead	Trommelfell n <Musik>	peau f de tambour	trommelvel n
D 195	drum membrane	Trommelfell n <Ohr>	tympan m	trommelvlies n
D 196	drum roll	Trommelwirbel m	roulement m de tambour	tromgeroffel n
D 197	drumsticks	Trommelstöcke mpl	baguettes fpl de tambour	trommelstokken pl
D 198	dry friction damping	trockene Reibungsdämpfung f	frottement m amortisseur sec	verlies n door droge wrijving
D 199	dual amplification	Reflexverstärkung f	amplification f de réflexion	reflexversterking
D 200	dual analogue	duale Analogie f	duale analogue m	duale analogie
D 201	dual cone loudspeaker, duplex loudspeaker	Doppelkonuslautsprecher m	haut-parleur m à double cône	dubbelconusluidspreker
D 202	dual diversity receiver	Zweikanalempfänger m	récepteur m bicanal	tweekanaalsontvanger
D 203	dual modulation	Doppelmodulation f	double modulation f	dubbele modulatie
D 204	dual network	duales Netzwerk n, duale Schaltung f	double réseau m	duaal netwerk n
D 205	dubbing	Einblendung f	enchainement m	geluidmenging, overschrijving
	dulcimer	s. C 64		
D 206	dull	dumpf	sourd, mat <son>	dof
D 207	dullness	Dumpfheit f	matité f <son>	dofheid

D 208	dumb	stumm	muet	stom, sprakeloos
D 209	duo-cone loudspeaker	Doppelkonuslautsprecher *m*	haut-parleur *m* à double cône	dubbelconusluidspreker
D 210	duodecimo	Duodezime *f*	duodécime *f*	duodeciem
D 211	dupe	Kopie *f*	double *m*, copie *f*	duplikaat *n*
D 212	duplet, couplet	Duole *f*	doublet *m*, couplet *m*	duool
	duplex loudspeaker	*s.* D 201		
D 213	duration allowance	Dauerzuschlag *m*, Dauer-zuschuß *m*	supplément *m* permanent	permanente toelage
D 214	duration of exposure	Einwirkungsdauer *f*	durée *f* d'exposition	expositietijd
D 215	duration of shock pulse	Impulsdauer *f*	durée *f* d'impulsion	pulstijd
D 216	dust noise	Staubrauschen *n*	bruit *m* de poussière (souillures)	ruis door stofdeeltjes
D 217	dwarf wave	Zwergwelle *f*	onde *f* minime	uiterst kleine golf
D 218	dying away	Ausschwingen *n*, Abklingen *n*, Verhallen *n*	évanouissement *m*	wegsterven
	dying oscillation	*s.* D 27		
D 219	dying-out constant ‹sl›	Ausschwingzeit *f*	constante *f* décrémentielle	uitklinktijd
D 220	dying-out transient	Ausschwingvorgang *m*	transit *m* d'évanouissement	voorbijgaand verschijnsel *n*
D 221	dynamic earphone	Dynamikkopfhörer *m*	écouteurs *mpl* dynamiques	elektrodynamische telefoon
D 222	dynamic range, DR	Lautstärkeumfang *m*, Dynamikbereich *m*	plage *f* de puissance acoustique	dynamiekspan
D 223	dynamic range expansion	Dynamikerweiterung *f*	expansion *f* de volume sonore	dynamiekverwijding
D 224	dynamic range of sound	Klangintensitätsbereich *m*	plage *f* d'intensité sonore	dynamiekspan van het geluid

E

E 1	ear cap, ear piece	Hör[er]muschel *f*	cape *f* (pavillon *m*, coquille *f*) d'écouteur	oorkapsel *n*
E 2	ear clip	Ohrbügel *m*	clip *m* d'oreille	oorbeugel
E 3	ear defender	Gehörschützer *m*	protecteur *m* d'ouïe	gehoorbeschermer
E 4	eardrum, tympanum	Trommelfell *n* ‹Ohr›	tympan *m*	trommelvlies *n*
E 5	ear muff	Ohrschutz *m*	protecteur *m* d'oreilles	oorkap
E 6	ear noises	Ohrensausen *n*	bourdonnements *mpl* d'oreilles	oorsuizen
E 7	earphone	Einsteckkopfhörer *m*	écouteur *m* à olive	oortelefoon
E 8	earphone coupler	Hörerkoppler *m*	accouplement *m* d'écouteurs	koppelstuk *n*
	ear piece	*s.* E 1		
E 9	earplug	Ohrstöpsel *m*	olive *f* auriculaire	oordopje *n*
E 10	ear-response characteristics	Ohrempfindlichkeitskenn-linie *f*	courbe *f* de sensibilité de l'ouïe	oorgevoeligheidskromme
E 11	earsplitting	schmerzhaft laut	faisant mal aux oreilles	oorverdovend
	ear trumpet	*s.* H 49		
E 12	echo, resound	widerhallen, zurückhallen	avoir de l'écho, résonner	weerklinken
E 13	echo	Echo *n*	écho *m*	echo
E 14	echo-attenuation measuring set	Echodämpfungsmesser *m*	échomètre *m* comparateur	nagalmmeetapparaat *n*
E 15	echo cancellation	Echokompensation *f*	compensation *f* d'écho	nagalmcompensatie
	echo depth sounding	*s.* A 70		
E 16	echo effect	Echowirkung *f*, Echoeffekt *m*	effet *m* d'écho	echo-effect
E 17	echoic	zum Echo gehörend	appartenant à l'écho	bij nagalm behorend
	echo killer ‹sl›	*s.* E 28		
E 18	echo killing ‹sl›	Echounterdrückung *f*	blocage *m* d'écho	nagalmomderdrukking
E 19	echolocation	Echoortung *f*	localisation *f* d'écho	plaatsbepaling door echo's
E 20	echo microphone	Echomikrofon *n*, Echo-effektmikrofon *n*	microphone *m* d'écho	nagalmmicrofoon
E 21	echo organ	Echowerk *n* ‹Orgel›	jeu *m* d'écho	echowerk *n*
E 22	echo path	Echoweg *m*, Echopfad *m*	circuit *m* d'écho	echopad *n*
E 23	echo pattern	Echobild *n*, Reflekto-gramm *n*	image *f* écho	reflectogram *n*
E 24	echo printing	Echokopie *f*, Echoüber-sprechen *n*	écho *m* diaphonie	echo-overspraak
E 25	echo ranging	Echoortung *f*	écho-localisation *f*	afstandsbepaling door echo's
E 26	echo sounder	Behmlot *n*	sonde *f* à écho	echolood *n*
	echo sounding	*s.* A 70		
E 27	echo splitting	Echospaltung *f*	division *f* d'écho	splitsing van echo's
E 28	echo suppressor, echo killer ‹sl›	Echosperre *f*	blocage *m* (suppresseur *m*) d'écho	echo-onderdrukker
E 29	echo time	Echolaufzeit *f*	temps *m* de propagation d'écho	looptijd van de echo
E 30	echo wave, back wave, reflected wave	Echowelle *f*, reflektierte Welle *f*	onde *f* réfléchie (d'écho)	teruggekaatste golf
E 31	eddy flow	Wirbelströmung *f*	écoulement *m* tourbillon-naire (de Foucault)	wervelstroming
E 32	eddy formation	Wirbelbildung *f*	formation *f* de tourbillons	wervelvorming
E 33	edge damping	Randdämpfung *f*	amortissement *m* en bordure	randdemping
E 34	edge tone	Randton *m*	son *m* en bordure	wigtoon
	effect channel	*s.* E 36		
	effective acoustic centre	*s.* A 56		
E 35	effect loudspeaker	Effektlautsprecher *m*	haut-parleur *m* à effet	effectluidspreker
E 36	effect train (way), effect channel	Effektkanal *m*	canal *m* à effet	effectenkanaal *n*
E 37	eigentone	Eigenton *m*	son *m* propre	eigentoon
E 38	eject key	Auswerftaste *f* ‹Kassetten-gerät›	touche *f* d'éjection	uitwerptoets
E 39	electroacoustical reciprocity theorem	elektroakustischer Rezi-prozitätssatz *m*	théorème *m* acoustique de réciprocité	theorema *n* van elektro-akoestische reciprociteit
E 40	electroacoustical transducer	elektroakustischer Wandler *m*	transducteur *m* électro-acoustique	elektroakoestische trans-ducent

E 41	electroacoustic coupling impedance, electroacoustic force factor	elektroakustischer Wandlerkoeffizient m	coefficient m de couplage acoustique	elektroakoestische-overdrachtsfactor
E 42	electroacoustic response	elektroakustische Empfindlichkeitskurve f	courbe f de réponse électroacoustique	elektroakoestische responsie
E 43	electroacoustics	Elektroakustik f	électro-acoustique f	elektroakoestiek
E 44	electrodynamic microphone, moving-conductor microphone	dynamisches Mikrofon n	microphone m électrodynamique (à ruban)	elektrodynamische microfoon
E 45	electromagnetic microphone	magnetisches Mikrofon n	microphone m électromagnétique	elektromagnetische microfoon
E 46	electromechanical coupling factor	elektromechanischer Kopplungsfaktor m	facteur m de couplage électromécanique	elektromechanische-koppelfactor
E 47	electromechanical coupling impedance, electromechanical force factor	elektromechanischer Wandlerkoeffizient m	coefficient m de couplage électromécanique	elektromechanische-overdrachtsfactor
E 48	electrophonic effect	elektrophonischer Effekt m	effet m électrophonique	elektrofonisch effect n
E 49	electrostatic actuator	elektrostatischer Anreger m	actuateur m électrostatique	elektrostatische aandrijver
	electrostatic microphone	s. C 9		
E 50	electrostriction	Elektrostriktion f	électrostriction f	elektrostrictie
E 51	electrostrictive	elektrostriktiv	électrostrictif	elektrostrictief
E 52	elimination of standing waves	Stehwellenentzerrung f, Unterdrückung f stehender Wellen	élimination f de l'ondes stationnaires	onderdrukking van staande golven
E 53	ellipsoidal wave	elliptische Welle f, Ellipsoidwelle f	onde f elliptique	ellipsoïdale golf
E 54	elliptically polarized sound wave	Schallwelle f mit elliptischer Polarisation	onde f sonore à polarisation elliptique	geluidgolf met elliptische polarisatie
E 55	embossing stylus	Eindrückstift m	stylet m d'embossage	beitel
E 56	emergent ray, issuing ray	austretender (ausfallender) Strahl m	rayon m émergent	uittredende straal
E 57	emission	Strahlung f, Ausstrahlung f, Aussendung f	émission f	emissie
E 58	emission of waves	Wellenaussendung f	émission f d'onde	uitstraling van golven
E 59	emphasis	Anhebung f, Betonung f	renforcement m	nadruk
E 60	emphasizer	Anheber m	renforceur m	uitlichter
E 61	energy density	Energiedichte f	densité f énergétique	energiedichtheid
E 62	energy efficiency	energetischer Wirkungsgrad m	rendement m énergétique	energetisch rendement n
E 63	enneaphony	Neunkanalübertragung f	enneaphonie f	enneafonie
E 64	entrance	Eingang m, Einführung f <Gerät>, Einsatz m <Musik>	entrée f, ouverture f	ingang
E 65	envelope	Hülle f, Hüllkurve f	enveloppe f	omhullende
E 66	envelope distortion	Modulationsverzerrung f	distorsion f de l'enveloppante	vervorming van de omhullende
E 67	epiotic	in Ohrnähe	à proximité d'oreille	vlak bij het oor
E 68	equalization for pressure increase	Druckstauentzerrung f	compensation f d'accroissement de pression	egalisatie bij drukopbouw
E 69	equal-loudness contours	Kurven fpl gleicher Lautstärke	courbes fpl d'égalité de puissance	krommen pl van gelijke luidheid
E 70	equally tempered scale	gleichmäßig temperierte Stimmung f	tonalité f uniformément tempérée	getempereerde toonladder
E 71	equally tuned	gleichgestimmt	accordé à l'unisson	gelijk gestemd
E 72	equal-noisiness contours	Kurven fpl gleicher Lärmlästigkeit	courbes fpl d'égalité de bruit	krommen pl van gelijke lawaaibelasting
E 73	equal-tempered	gleichtemperiert <Musik>	uniformément tempéré	getempereerd
E 74	equivalent absorption	äquivalente Absorption f	absorption f équivalente	equivalente absorptie
E 75	equivalent absorption area	äquivalente Absorptionsfläche f	surface f d'absorption équivalente	equivalent absorberend oppervlak n
E 76	equivalent acoustics	Ersatzschalleffekte mpl	effets mpl acoustiques équivalents	speciaal opgenomen vervangingsgeluiden npl
E 77	equivalent articulation loss	Ersatzdämpfung f <für Sprachverständlichkeit>	amortissement m équivalent d'intelligibilité	equivalent verlies n van spraakverstaanbaarheid
E 78	equivalent continuous sound level, equivalent loudness level	äquivalenter Dauerlärmpegel m (Dauerschallpegel m)	niveau m continu de son équivalent	equivalent geluidpeil n, equivalent luidheidspeil n
E 79	equivalent noise fourpole <twoport>	äquivalenter Rauschvierpol m <Rauschzweitor>	quadripôle m équivalent de bruit <double porte>	equivalente ruisvierpool
E 80	equivalent noise level	Ersatzstörlautstärkepegel m	niveau m équivalent de bruit	equivalent geruispeil n
E 81	equivalent noise pressure	äquivalenter Rauschdruck m (Druck m des Rauschens)	pression f de bruit équivalente	equivalente geruisdruk
E 82	equivalent noise source diagram	Rauschquellenersatzschaltbild n	source f de bruit équivalente	schema n van equivalente ruisbronnen
E 83	equivalent noise twoport	äquivalentes Rauschzweitor m <Vierpol>	quadripôle m à bruit équivalent	equivalente ruisvierpool
E 84	equivalent piston	Äquivalentkolben m	piston m équivalent	equivalente trilplaat
E 85	equivalent thres old sound pressure level	äquivalenter Schwellen-Schalldruckpegel m	niveau m équivalent de seuil de pression sonore	equivalente geluiddrukpeildrempel
E 86	erasing preventing device	Löschsperre f <Tonband>	verrouillage m d'effacement	beveiliging tegen ongewild wissen
E 87	erratic	sprunghaft	erratique	wild
	ESP	s. E 104		
	euphonium	s. B 51		
E 88	Eustachian tube	Eustachische Röhre f	trompe f d'Eustache	buis van Eustachius
E 89	even harmonic	geradzahlige Oberwelle f	harmoniques fpl d'ordre pair	even harmonische
E 90	Ewing test	Hörprüfung f für Kinder <mit vertrauten Lauten>	test m auditif pour enfants	gehoortest voor kinderen
E 91	excitation stimulus	Anregung f	excitation f	stimulus
E 92	excite	erregen, anregen	exciter	prikkelen
E 93	exhaust silencer	Schalldämpfer m	pot m d'échappement silencieux <d'un moteur à explosion>	knalpot

E 94	expanded speech	gedehnte Sprache *f*	parole *f* expandée	verwijde spraak
E 95	expandor	Dynamikdehner *m*	expanseur *m*	dynamiekverwijder
	expansion	*s.* C 188		
E 96	exponential horn	Exponentialschalltrichter *m*	pavillon *m* exponentiel	exponentiële hoorn
E 97	exponential-horn loud-speaker	Exponentialtrichter-Laut-sprecher *m*	haut-parleur *m* à pavillon exponentiel	luidspreker met exponen-tiële hoorn
E 98	extension organ	Multiplexorgel *f*	orgue *m* multiplex	multiplexorgel *n*
E 99	external loudspeaker	Außenlautsprecher *m*	haut-parleur *m* externe	losse luidspreker
E 100	external modulation	Fremdmodulation *f*	modulation *f* externe	externe modulatie
E 101	extra-axial beam	außeraxialer Strahl *m*	rayon *m* excentrique (non axial)	straal buiten de as
E 102	extraneous field	äußeres Feld *n*, Fremdfeld *n*	champ *m* extérieur	uitwendig veld *n*
E 103	extraneous noise	Fremdgeräusch *n*	bruit *m* extérieur	van buiten komende ruis
E 104	extrasensory perception, ESP	außersinnliche Wahr-nehmung *f*	perception *f* extra-sensorielle	buitenzintuigelijke waarne-ming
E 105	extremities	Enden *npl* ‹des Bandes›	extrémités *fpl*, fins *fpl* de bande	grenzen
E 106	Eyring coefficient	Eyringscher Schallabsorp-tionsexponent *m*	coefficient *m* d'absorption du son de Eyring	absorptiecoëfficiënt volgens Eyring

F

F 1	face-to-face phone, see-as-you-talk phone ‹US›	Bildtelefon *n*	téléphone *m* audio-visuel	televisiefoon
F 2	fade	schwinden, Schwund haben	s'évanouir, avoir du fading	verdwijnen
F 3	fade	Schwund *m*	fading *m*	geleidelijke overgang
F 4	fade down	abschwächen	affaiblir	geleidelijk verzwakken
F 5	fade in	einblenden	enchaîner peu à peu	langzaam opkomen
F 6	fade out	ausblenden	affaiblir jusqu'à l'extinction	langzaam verdwijnen
F 7	fader	veränderliches Dämpfungs-glied *n*, Blendregler *m*	régleur *m* d'affaiblissement	overgangsregelaar
F 8	fade up	verstärken	renforcer, amplifier	geleidelijk versterken
F 9	fading effect	Schwundeffekt *m*, Schwund-erscheinung *f*	effet *m* de fading, fading *m*	effect *n* van geleidelijke ver-anderingen
F 10	fading frequency	Schwundhäufigkeit *f*, Schwundfrequenz *f*	fréquence *f* de fading	frequentie van geleidelijke veranderingen
F 11	fading hollow	Wellenschwundloch *n*	trou *m* de fading	uitdovingsgat *n*
F 12	fading-numbers test	Zahlenflüster-Hörprüfung *f*	test *m* de nombres murmurés	gehoortest met gefluisterde getallen
F 13	faithful reproduction	originaltreue Wiedergabe *f*	reproduction *f* d'absolue fidélité	natuurgetrouwe weergave
F 14	fan out	[sich] fächerförmig aus-breiten	diffuser en éventail	uitwaaieren
F 15	far field	Fernfeld *n*	champ *m* éloigné	verre veld *n*
F 16	fast responding	schnell ansprechend	à réponse rapide	met snelle responsie
F 17	fast response	Dynamik „Fast", Dynamik „Schnell" ‹beim Schall-pegelmesser›	dynamique *f* rapide	snelle responsie
	fathometer	*s.* A 69		
F 18	feedback	rückkoppeln	coupler en retour	terugkoppelen
F 19	feedback, back-coupling	Rückkopplung *f*	réaction *f*, couplage *m* en retour	terugkoppeling
F 20	feedback circuit	Rückkopplungsschaltung *f*	circuit *m* de réaction	terugkoppelschakeling
F 21	feedback circuit	Einspielkreis *m*	circuit *m* de réaction	terugkoppelingskring
F 22	feigned deafness	simulierte Taubheit *f* (Schwerhörigkeit *f*)	surdité *f* simulée	oostindische doofheid
	FF	*s.* F 71		
F 23	fidelity of response	Wiedergabetreue *f*	fidélité *f* de réponse	getrouwheid van weergave
F 24	field-to-noise ratio	Störfeldabstand *m*	rapport *m* perturbation-champ	relatief ruispeil *n*
F 25	fife	Querpfeife *f*	petite flûte *f*	dwarspijp
F 26	figured bass, basso continuo	Generalbaß *m*	basse-contre *f*, continuo *m*, basse *f* fondamentale	becijferde bas
F 27	figured melody	Figuralmelodie *f*	thème *m* conducteur	figuraalmelodie
F 28	figure of merit of reception	Empfangsgüte *f*	qualité *f* de réception	ontvangstkwaliteit
F 29	figure of performance	Sonar-Gütezahl *f*	coefficient *m* de qualité	prestatiegetal *n* ‹van een sonar›
F 30	filter	sieben	filtrer	filteren
F 31	filter	Sieb *n*	filtre *m*	filter *n*
F 32	filter choke	Siebdrossel *f*	bobine *f* d'arrêt (de choke)	smoorspoel
F 33	filter coupling	Filterkopplung *f*	couplage *m* de filtre	filterkoppeling
F 34	filter ladder	Filterkette *f*	chaîne *f* de filtrage	filterketen
F 35	filter out	ausfiltern, aussieben	filtrer, éliminer	uitfilteren
F 36	final amplifier, terminal amplifier	Endverstärker *m*	amplificateur *m* final	eindversterker
F 37	final chain	Meßstrecke *f* ‹zwischen Ausgangsverstärker und entferntestem Meßgerät›	chaîne *f* de mesure	meetketen
F 38	flanking channel	Nachbarkanal *m*, Flanken-kanal *m*	canal *m* voisin (contigu)	aangrenzend kanaal *n*
F 39	flanking transmission	Flankenübertragung *f*, Nebenwegübertragung *f*	transmission *f* latérale	flankerende overdracht
F 40	flared throat	aufgekelchter Trichterhals *m*	pavillon *m* évasé	verwijde hals
F 41	flare factor	Krümmungsexponent *m* des Lautsprechertrichters	facteur *m* de cintrage	welvingsfactor
F 42	flat frequency response	linearer Frequenzgang *m*	réponse *f* en fréquence linéaire	vlakke frequentiekarakteris-tiek
F 43	flat loudspeaker, pancake loudspeaker ‹sl›, wafer loudspeaker	Flachlautsprecher *m*	haut-parleur *m* plat	platte luidspreker
F 44	flat room	Flachraum *m*	salle *f* basse	ondiepe kamer

F 45	flat-topped resonance crest	abgeflachte Resonanzspitze f	courbe f de résonance aplatie	afgeplatte resonantiepiek
F 46	flexible diaphragm	bewegliche Membran f	membrane f mobile	buigzaam membraan n
F 47	flexural natural frequency	Biegeeigenfrequenz f	fréquence f propre de flexion	eigenfrequentie van de buigtrilling
F 48	flexural resonator	Biegeschwinger m	piézo-résonateur m longitudinal	buigtriller
F 49	flexural wave	Biegewelle f	onde f de flexion	buiggolf
F 50	flow resistance	Strömungsresistanz f	résistance f de flux, résistance f à l'écoulement	stromingsweerstand
F 51	flow resistivity	längenspezifische Strömungsresistanz f	résistivité f à l'écoulement	specifieke stromingsweerstand
	flue pipe	s. L 1		
F 52	flugelhorn, saxhorn, vocal horn	Flügelhorn n	trompette f d'harmonie, saxhorn m, piston m	saxhoorn
F 53	flute	Flöte f	flûte f	fluit
F 54	flutter	Schwankung f der Bandgeschwindigkeit	variation de la vitesse de translation de la bande	zwevingen pl
F 55	flutter and wow	Gleichlaufschwankung f	flottement m	zwevingen pl en janken n
F 56	flutter echo	Flatterecho n, Schätterechon	flottement m d'écho	meervoudige echo
F 57	foam rubber ear pad	Schaumgummimuschel f <Kopfhörer>	cape f d'écouteur en mousse de caoutchouc	schuimrubber oorkap
F 58	folded-horn loudspeaker	Faltenlautsprecher m	haut-parleur m à membrane plissée	luidspreker met gevouwen hoorn
F 59	following microphone	bewegliches Mikrofon n	microphone m mobile	beweeglijke microfoon
	footfall sound	s. I 15		
F 60	forced oscillation, forced vibration	erzwungene Schwingung f	oscillations fpl forcées	gedwongen trilling
F 61	force factor	Kraftfaktor m	facteur m de force	krachtfactor
F 62	fork beat	Stimmgabelschwebung f	vibration f de diapason	vorktrilling
F 63	formant	Formant m	formant m	formant
F 64	formant-coding speech analyser	formantkodierender Sprachanalysator m	analyseur m de parole codant les formants	formantcoderende spraakanalysator
F 65	form of oscillation	Schwingungsform f	forme f d'oscillation	trillingsvorm
	forward facing speaker	s. F 100		
F 66	four-channel magnetic head	Vierkanalmagnetkopf m	tête f magnétique à quatre canaux	weergeefkop voor vier kanalen
F 67	four-channel power amplifier	Vierkanalhauptverstärker m	amplificateur m de puissance quadricanal	vierkanaals eindversterker
F 68	four-part	vierstimmig	à quatre parties (voix)	vierstemmig
	four-terminal network	s. T 168		
F 69	fractional detuning	Feinverstimmung f	désaccord m partiel	gedeeltelijke verstemming
F 70	fractional pitch	Teilschritt m	fraction f de pas	toontrede
	freedom from noise	s. A 2		
F 71	free field, FF, free sound field	freies Schallfeld n	champ m acoustique libre	vrije veld n
F 72	free-field conditions	Freifeldbedingungen fpl	conditions fpl de champ libre	omstandigheden pl van een onbegrensde ruimte
F 73	free-field current sensitivity	Freifeldstromübertragungsfaktor m	sensitivité f de courant de champ libre	stroomgevoeligheid in het vrije veld
F 74	free-field room	Freifeldraum m, reflexionsfreier Raum m	chambre f à champ libre	reflectieloze ruimte
F 75	free-field sensitivity	Freifeldempfindlichkeit f	sensitivité f en champ libre	gevoeligheid in het vrije veld
F 76	free-field voltage response (sensitivity)	Freifeldspannungsübertragungsfaktor m	réponse f en tension en champ libre	spanningsgevoeligheid in het vrije veld
F 77	free impedance	elektrische Eingangsimpedanz f bei frei schwingendem System	impédance f d'entrée d'un système libre	ingangsimpedantie bij onbelaste uitgang
F 78	free oscillation	freie Schwingung f	oscillation f libre	vrije trilling
F 79	free progressive wave	freie fortschreitende Welle f	onde f de propagation libre	vrije lopende golf
	free sound field	s. F 71		
F 80	free space pattern	Freiraumdiagramm n <Strahlung>	diagramme m en espace libre	stralingsdiagram n in de vrije ruimte
F 81	free space propagation	Freiraumausbreitung f	propagation f en espace libre	voortplanting in de vrije ruimte
	French horn	s. C 149		
F 82	frequency characteristic, frequency response	Frequenzkennlinie f, Frequenzkurve f, Frequenzgang m	courbe f de réponse en fréquence	frequentiekarakteristiek
F 83	frequency detector, discriminator	Frequenzdiskriminator m	discriminateur m de fréquence	discriminator
F 84	frequency discrimination	Frequenzauflösung f	discrimination f de fréquence	frequentieonderscheid n
F 85	frequency drift	Frequenzauswanderung f	glissement m de fréquence	verloop n van de frequentie
F 86	frequency equalization	Frequenzgangentzerrung f	égalisation f de fréquences	frequentie-egalisatie
F 87	frequency error	Frequenzabweichung f	erreur f de fréquence	frequentieafwijking
F 88	frequency jumping	Frequenzsprung m	saut m de fréquence	frequentiesprong
F 89	frequency level	Frequenzpegel m	niveau m de fréquence	frequentiepeil n
F 90	frequency modulation	Frequenzmodulation f	modulation f de fréquence	frequentiemodulatie
F 91	frequency of attenuation	Frequenzgang m der Dämpfung	fréquence f d'atténuation	frequentie waarbij verzwakking optreedt
F 92	frequency of oscillations	Schwingungsfrequenz f	fréquence f des oscillations	trillingsfrequentie
F 93	frequency of recurrence	Wiederholungsfrequenz f	fréquence f de répétition	herhalingsfrequentie
F 94	frequency resolution	Frequenzauflösung f	résolution f de fréquence	frequentieresolutie
	frequency response	s. F 82		
F 95	frequency spacing	Frequenzabstand m	éloignement m de fréquence	frequentie-tussenruimte
F 96	frequency stability	Frequenzkonstanz f	stabilité f de fréquence	frequentiestabiliteit
F 97	frequency standard	Frequenznormal n	fréquence f de référence, standard m de fréquence	standaardfrequentie
F 98	frequency weighting	Frequenzbewertung f	pondération f de fréquence	frequentieafhankelijke weging
F 99	fringe area	Randgebiet n	zone f de bordure	randgebied n
F 100	front mounted [loud-] speaker, forward facing speaker	Frontlautsprecher m	haut-parleur m frontal	frontale luidspreker

F 101	**fry**	Grundgeräusch *n*, Knistergeräusch *n* ‹Mikrofon›	bruit *m* de fond (friture)	ritselen
F 102	**frying**	Knistern *n*	crépitement *m*	geritsel *n*
F 103	**full echo suppressor**	Vollechosperre *f*	suppresseur *m* total d'écho	totale echo-onderdrukker
F 104	**fullness of tone**	Klangfülle *f*	plénitude *f* de tonalité	volheid van klank
F 105	**functional communication profile**	funktionelles Kommunikationsprofil *n*	profil *m* de communication fonctionnel	functioneel communicatieprofiel *n*
F 106	**functional deafness**	funktionelle Taubheit *f*	surdité *f* fonctionnelle	functionele doofheid
F 107	**fundamental frequency**	Grundfrequenz *f*	fréquence *f* fondamentale	grondfrequentie
F 108	**fundamental-frequency attenuation**	Grundfrequenzdämpfung *f*	atténuation *f* de la fréquence fondamentale	verzwakking van de grondfrequentie
F 109	**fundamental harmonic, fundamental oscillation**	Grundschwingung *f*	harmonique *f* fondamentale	grondtrilling
F 110	**fundamental mode of vibration**	Grundschwingungsmode *f*	mode *m* fondamental de vibration	fundamentele trillingswijze
	fundamental oscillation	*s.* F 109		
F 111	**fundamental tone**	Grundton *m*	ton *m* fondamental	grondtoon
F 112	**fundamental wave**	Grundwelle *f*	onde *f* fondamentale	fundamentele golf
F 113	**fundamental wavelength**	Grundwellenlänge *f*	longueur *f* d'onde fondamentale	fundamentele golflengte

G

G 1	**gain**	Verstärkung *f*, Verstärkungsgrad *m*	amplification *f*, gain *m*	versterking
G 2	**gain characteristics**	Verstärkungskennlinie *f*	caractéristique *f* d'amplification (de gain)	versterkingskarakteristieken *pl*
G 3	**galactic noise**	galaktisches Rauschen *n*	bruit *m* galactique	ruis uit de Melkweg
G 4	**Galton's whistle**	Galton-Pfeife *f*	tube *m* de Galton	fluit van Galton
G 5	**galvanometer recorder**	dynamischer Lichthahn *m*	galvanomètre *m* enregistreur à miroir	galvanometerschrijver
G 6	**gamut**	diatonische Tonleiter *f*	gamme *f* diatonique	diatonische toonladder
G 7	**gap alignment**	Magnetkopfjustierung *f*	ajustement *m* de tête magnétique	justering van de koppen
G 8	**gap length**	Spaltbreite *f* ‹beim Magnetkopf›	largeur *f* de gap, largeur d'entrefer	spleetlengte
G 9	**gap scatter**	Spaltlagenstreuung *f*	dispersion *f* de localisation de gap	ongelijke positie van de spleten
G 10	**Gaussian noise**	Gaußsches Rauschen *n*	bruit *m* gaussien	Gaussische ruis
G 11	**geoacoustics**	Geoakustik *f*	géoacoustique *f*	geoakoestiek
G 12	**geometrical attenuation**	geometrische Dämpfung *f*	atténuation *f* géométrique	geometrische verzwakking
G 13	**geometrical moment of inertia**	Flächenträgheitsmoment *n*	moment *m* géométrique d'inertie	geometrisch traagheidsmoment *n*
G 14	**geophone**	Geofon *n*	géophone *m*	geofoon
	German flute	*s.* C 230		
G 15	**gipsies' scale**	Zigeuner[ton]leiter *f*	gamme *f* tzigane	Zigeuner-toonladder
G 16	**glottis** ┌**phone**	Stimmritze *f*	glotte *f*	stemspleet
	glow discharge micro-	*s.* C 23		
G 17	**glow lamp sound recording**	Glimmlampenschallaufnahme *f*	enregistrement *m* avec lampe au néon	geluidregistratie met een glimlamp
G 18	**gobo**	Schluckschirm *m*, Schallschluckschirm *m*	écran *m* atténuateur	geluiddempend scherm
G 19	**graded filter**	abgestuftes Filter *n*	filtre *m* gradué	getrapt filter *n*
G 20	**gradient microphone**	Gradientmikrofon *n*	microphone *m* de gradient	gradiëntmicrofoon
G 21	**grand piano**	Flügel *m*	piano *m* à queue	vleugel
	granule microphone	*s.* C 13 ┌**fon**›		
G 22	**granules**	Kohlekörner *npl* ‹im Mikro-	grenaille *f*	koolkorrels *pl*
G 23	**grid hum**	Gitterbrumm *m*	ronflement *m* de la grille	roosterbrom
G 24	**grid suppressor**	Gittervorwiderstand *m* zur Unterdrückung störender Schwingungen	suppresseur *m* de grille	stopweerstand
G 25	**grille cloth**	Lautsprecherbespannstoff *m*	tenture *f* de haut-parleur	luidsprekerdoek *n*
G 26	**grille-type microphone**	Gittermikrofon *n*	microphone *m* à grille	roostermicrofoon
G 27	**groove**	Rille *f* ‹Schallplatte›	sillon *m*, sommier *m*	groef
G 28	**groove shape**	Rillenform *f*	forme *f* de sillon	groefdoorsnede
G 29	**groove spacing**	Füllgrad *m* der Schallplatte	degré *m* de remplissage	efficiënte groefverdeling
G 30	**ground back-scatter**	Bodenrückstreuung *f*	degré *m* de dispersion en retour par le sol	terugverstrooiing van de bodem
G 31	**ground noise**	Eigenrauschen *n*, Grundgeräusch *n*	bruit *m* de fond	eigenruis
G 32	**ground ray**	Bodenwelle *f*	onde *f* de sol	grondgolf
G 33	**ground reflection factor**	Bodenreflexionsfaktor *m*	facteur *m* de réflexion au sol	reflectiecoëfficiënt van de bodem
G 34	**group delay time**	Gruppenlaufzeit *f*	temps *m* de propagation de groupe	looptijd van de groef
G 35	**group frequency**	Gruppenfrequenz *f*, Wellenzugfrequenz *f*	fréquence *f* de train d'ondes	groepfrequentie
G 36	**group hearing aid**	Gruppenhörhilfe *f*, Hörleiste *f*	aide *f* acoustique commune (à un groupe)	gemeenschappelijke hoorhulp
G 37	**grouping**	Füllschritt *m*	groupage *m*	groefdistributie
G 38	**group velocity**	Gruppengeschwindigkeit *f*	vélocité *f* de groupe	groepsnelheid
G 39	**gut-string**	Darmsaite *f*	corde *f* en boyau	darmsnaar
G 40	**guttural sound**	Kehllaut *m*	son *m* guttural	keelklank, gutturaal

H

	HAE	*s.* H 39		
H 1	**half-cycle**	Halbperiode *f*	demi-période *f*	halve periode
H 2	**half-power point**	Halbwertspunkt *m* ‹der Energie›	point *m* de demi-puissance	—3 dB-punt *n*, minus-driedecibel-punt *n*

	English	German	French	Dutch
H 3	half-step	Halbton *m*	demi-ton *m*	halve toon
H 4	half-wave	Halbwelle *f*	demi-onde *f*	halve golf
	hall noise	*s.* A 296		
H 5	hand-held microphone, hand microphone	Handmikrofon *n*	microphone *m* portatif	draagbare microfoon
H 6	hardly perceptible	kaum wahrnehmbar	à peine perceptible	nauwelijks waarneembaar
	hardness of hearing	*s.* A 196		
H 7	hard-of-hearing	schwerhörig	partiellement sourd (dur d'oreille)	hardhorend
	harmonic	*s.* H 16		
H 8	harmonic analysis	Fourier-Analyse *f*	analyse *f* de Fourier	harmonische analyse, Fourier-analyse
H 9	harmonic components	harmonische Teilschwingungen *fpl*	composantes *fpl* harmoniques	hogere harmonischen *pl*
H 10	harmonic content	Oberwellengehalt *m*	teneur *f* en harmoniques	gehalte *n* aan harmonischen
	harmonic distortion	*s.* D 151		
H 11	harmonic filter	Oberwellenfilter *n*	filtre *m* d'harmoniques	harmonischen-filter *n*
H 12	harmonic interference	Oberwellenstörung *f*	interférence *f* d'harmoniques	storing door harmonischen
H 13	harmonic minor	harmonische Molltonleiter *f*	gamme *f* mineure harmonique	harmonische toonladder
H 14	harmonic motion	sinusförmige Wellenbewegung *f*	mouvement *m* ondulatoire sinusoïdal	harmonische beweging
H 15	harmonic number, order of harmonics	Ordnungszahl *f* der Harmonischen	numéro *m* d'harmonique, ordre *m* des harmoniques	harmonisch ranggetal *n*
H 16	harmonic oscillation, harmonic	harmonische Schwingung *f*, Oberschwingung *f*, Oberwelle *f*	oscillation *f* harmonique	harmonische trilling
H 17	harmonic oscillator	Sinusoszillator *m*	oscillateur *m* sinusoïdal	harmonische oscillator
H 18	harmonic pipe	überblasene Pfeife *f*	flûte *f* vibrant en harmonique	harmoniek
H 19	harmonic ratio	Klirrdämpfung *f*	proportion *f* d'harmoniques	harmonische verhouding
H 20	harmonic series	harmonische Reihe *f*	série *f* (échelle *f*) harmonique	harmonische reeks
H 21	harmonic series of sounds	einfacher (harmonischer) Klang *m*	sons *mpl* harmoniques simples	reeks van harmonische klanken
H 22	harmonic series of sounds	Naturtonreihe *f*	sons *mpl* harmoniques naturels	harmonische klankenreeks
H 23	harmonics graduation	Harmonischenleiter *f*	échelle *f* harmonique	harmonische toonladder
H 24	harmonic suppressor	Oberwellensperrfilter *n*	suppresseur *m* d'harmoniques, piège *m* à harmoniques	onderdrukker van harmonischen
H 25	harmonic tone	harmonischer Teilton *m*, Harmonische *f*	ton *m* harmonique	harmonische toon
H 26	harmonic wave	harmonische Welle *f*	onde *f* harmonique	harmonische golf
H 27	harmony whistle	Zweiklangpfeife *f*	tuyau *m* à double ton	tweetonig fluitje *n*
H 28	harp	Harfe *f*	harpe *f*	harp
H 29	harpsichord	Cembalo *n*	cembalo *m*	clavecimbel *n*
H 30	harsh	hart <Ton>	dur	rauw
H 31	hawaiian guitar, ukulele	Hawaiigitarre *f*, Ukulele *n*	ukulele *m*, guitare *f* hawaiienne	ukulele
H 32	headphone, head receiver	Kopfhörer *m*	écouteur *m*	hoofdtelefoon
H 33	headphones	Doppelkopfhörer *m*	écouteurs *mpl*	hoofdtelefoon
	head receiver	*s.* H 32		
H 34	headset	Mikrofon-Kopfhörerkombination *f*, Sprechgeschirr *n* <sl>	microphone *m* combiné	hoofdstel
H 35	head stack	Mehrspurmagnetkopf *m*	tête *f* magnétique multiple	meersporenkop
H 36	head voice	Kopfstimme *f*	voix *f* de tête	kopstem
H 37	hearing	Hören *n*	ouïe *f*	gehoor *n*
H 38	hearing aid, deaf aid, audicle, acouphone	Hörhilfe *f*	aide *f* acoustique	hoortoestel *n*
H 39	hearing aid evaluation, HAE	Hörhilfenbestimmung *f*	évaluation *f* de l'aide acoustique nécessaire	aanmeten van een hoortoestel
H 40	hearing-aid glasses	Hörbrille *f*	monture *f* de lunettes à écouteurs	hoorbril
H 41	hearing conservation	Gehörerhaltung *f*, Erhaltung *f* des Hörvermögens	conservation *f* de l'ouïe	gehoorbescherming
H 42	hearing damage	Gehörschädigung *f*	lésion *f* de l'ouïe	gehoorbeschadiging
H 43	hearing defect	Gehörfehler *m*	défaut *m* auditif	gehoordefect *n*
H 44	hearing disability	Hörbeeinträchtigung *f*	diminution *f* de l'ouïe	gehoorstoornis
H 45	hearing handicap	Hörbehinderung *f*, Gehörbehinderung *f*	réduction *f* de l'acuité auditive	gehoorvermindering
H 46	hearing impairment	Gefährdung *f* des Gehörs	dégradation *f* de l'ouie	gehoorverzwakking
H 47	hearing level, HL	Hörpegel *m* <dBA über audiometrischem Nullpunkt>	niveau *m* d'audibilité	gehoorpeil *n*
H 48	hearing loss	Hörverlust *m*	perte *f* de l'ouïe	gehoorverlies *n*
	hearing sensitivity	*s.* A 306		
H 49	hearing tube, ear trumpet	Hörrohr *n*	cornet *m* acoustique	gehoortrechter
H 50	heavy modulation	starke Modulation *f*	modulation *f* profonde (puissante)	diepe modulatie
H 51	heel	Frosch *m* <Musik>	couac *m*	slof
H 52	helicon	Kontrabaßtuba *f*, Helikon *n*	hélicon *m*	helicon
H 53	Helmholtz double layer	Helmholtzsche Doppelschicht *f*	double couche *f* de Helmholtz	dubbellaag van Helmholtz
H 54	Helmholtz resonator	Helmholtz-Resonator *m*	résonateur *m* de Helmholtz	Helmholz-resonator
H 55	Hertzian waves	Hertzsche Wellen *fpl*	ondes *fpl* hertziennes	golven *pl* van Hertz
H 56	heterodyne	überlagern	hétérodyner	heterodyne
H 57	heterodyne principle	Überlagerungsprinzip *n*	principe *m* hétérodyne (de superposition)	heterodyne-principe *n*
H 58	heterodyne wave	Überlagerungswelle *f*	onde *f* hétérodyne	heterodynegolf
H 59	heterodyne wavemeter	Interferenzwellenmesser *m*	ondemètre *m* hétérodyne	heterodynegolfmeter
H 60	hewgag	Rassel *f*, Lärminstrument *n*	crécelle *f*	ratel

H 61	hexachord	große Sexte *f*	accord *m* de sixte	hexachord
H 62	high band	frequenzhöheres Band *n*	bande *f* de fréquence supérieure	hoge frequentieband
H 63	high-compliance woofer <US>	Tieftonlautsprecher *m* mit guter akustischer Federung	haut-parleur *m* de graves de haute compliance	lagetonenluidspreker met hoge compliantie
H 64	high cut-off frequency	obere Grenzfrequenz *f*	limite *f* supérieure de fréquence	bovenste grensfrequentie
H 65	high damping	starke Dämpfung *f*	fort amortissement *m*	sterke demping
H 66	high fence <US>	obere Grenze *f* der Schwerhörigkeit <zur Taubheit>	limite *f* supérieure de la dureté d'oreille	bovenste begrenzing
H 67	high-fidelity microphone	Mikrofon *n* hoher Güte	microphone *m* de haute fidélité	microfoon met natuurgetrouwe weergave
H 68	high-fidelity reproduction	originalgetreue Wiedergabe *f*	reproduction *f* de haute fidélité	natuurgetrouwe weergave
H 69	high-frequency absorber	Schallschlucker *m* für hohe Frequenzen, Höhenschlucker *m*	absorbant *m* pour hautes fréquences	geluiddemper voor hoge frequenties
H 70	high-frequency emphasis	Höhenanhebung *f*	amplification *f* des aiguës	nadruk op de hogere frequenties
H 71	high-frequency peaking	Anhebung *f* der hohen Frequenzen	renforcement *m* des aiguës	versterking van de hoge frequenties
H 72	high-pass filter	Hochpaßfilter *n*	filtre *m* passe-haut	hoogdoorlaatfilter *n*
H 73	high-pitched	hellklingend, helltönend	à son clair	hoog klinkend
H 74	high-pitched tone	hoher Ton *m*	ton *m* haut	hoge toon
H 75	high-pitched voice	hohe Stimme *f*, Fistelstimme *f*	voix *f* haute, fausset *m*	hoge stem
H 76	hill and dale recording	vertikale Aufzeichnung *f* <senkrecht zur Plattenoberfläche>, Tiefenschrift *f*	gravure *f* en profondeur	dieptreschrift *n*
H 77	hiss	Zischen *n*	sifflement *m*	gesis *n*
	hiss	*s. a.* W 33		
	HL	*s.* H 47		
H 78	hole	tote Zone *f*	zone *f* sourde (morte)	gat *n*
H 79	hollow, wave trough	Wellental *n*, Wellensenke *f*	creux *m* d'onde, creux de vague	golfdal *n*
H 80	hollowness	Dumpfheit *f* <des Tones>	surdité *f* <son>	holheid
H 81	hollow space	Hohlraum *m*	espace *m* creux, cavité *f*	holle ruimte
H 82	hollow-space resonator	Hohlraumresonator *m*	résonateur *m* à cavité	holteresonator
H 83	homophony	Einstimmigkeit *f* <Musik>	unisson *f*, homophonie *f*	éénstemmigheid
H 84	hood loudspeaker	Baldachinlautsprecher *m*	haut-parleur *m* couvert	baldakijnluidspreker
H 85	horizontal directivity	Horizontalrichtwirkung *f*	directivité *f* horizontale	horizontale bundeling
H 86	horn	Schalltrichter *m*	pavillon *m* acoustique	hoorn
H 87	hornless loudspeaker	trichterloser Lautsprecher *m*	haut-parleur *m* sans pavillon	luidspreker zonder hoorn
H 88	horn loudspeaker	Trichterlautsprecher *m*	haut-parleur *m* à pavillon	hoornluidspreker
H 89	hot-wire microphone, thermal microphone	Hitzdrahtmikrofon *n*	microphone *m* à fil chaud	thermomicrofoon
H 90	howl	heulen	hurler	huilen
H 91	howl	Heulton *m*	hurlement *m*	gehuil *n*
H 92	howler	Heuler *m*	hurleur *m*	huiler
H 93	howling	Selbsttönen *n* <Verstärker>	accrochage *m*	rondzingen
	howl round	*s.* A 78		
H 94	hum	brummen	ronfler	brommen
H 95	hum, ripple	Brumm *m*	ronflement *m*, ondulation *f*	brom
H 96	hum	Brummspannungsabstand *m*	coefficient *m* de ronflement	bromniveau *n*
H 97	hum bar	Brummstreifen *m* <Fernsehen>	barre *f* de ronflement	zwarte balk
H 98	hum component, ripple component	Brummkomponente *f*	composante *f* de ronflement	bromcomponent
H 99	hum eliminator	Entbrummer *m*	éliminateur *m* de ronflement	ontbrommer
H 100	hum filtering	Brummsiebung *f*	filtre *m* anti-ronfleur	bromfiltering
H 101	hum frequency, ripple frequency	Brummfrequenz *f*	fréquence *f* de ronflement	bromfrequentie
H 102	humming	Summen *n*, Brummen *n*	bourdonnement *m*, ronflement *m*	gebrom *n*
H 103	humming noise	Brummton *m*	ronflement *m*	gezoem *n*
H 104	humming tone	Summerton *m*	bourdonnement *m*	zoemtoon
H 105	hum pick-up	Brummeinkopplung *f*	couplage *m* de bourdonnement	het oppikken van brom
H 106	hum sidebands	Brummseitenbänder *npl*	bandes *fpl* latérales de ronflement	zijbanden *pl* door brom
H 107	hum trouble	Brummstörung *f*	trouble *m* de ronflement	bromstoornis
H 108	hunting horn	Hifthorn *n*, Jagdhorn *n*	cor *m* de chasse	jachthoorn
H 109	hurdy-gurdy	Leierkasten *m*, Drehorgel *f*	orgue *m* de Barbarie	draaiorgel
H 110	hydrodynamic oscillator	hydrodynamischer Schwinger *m*	oscillateur *m* hydrodynamique	hydrodynamische triller
H 111	hydrophone	Hydrofon *n*	hydrophone *m*	hydrofoon
H 112	hyper-cardioid response	Supernierencharakteristik *f*	caractéristique *f* en super-cardioïde	hypercardioïde-responsie
H 113	hypersynchronous	übersynchron	hypersynchrone	hypersynchroon

I

I 1	ideal transducer	passiver Übertrager *m* vollen Wirkungsgrades	transducteur *m* idéal	ideale transducent
I 2	identification	Identifikation *f*, Identifizierung *f*	identification *f*	identificatie
	IL	*s.* I 87		
I 3	ill-balanced	fehlerhaft symmetriert	mal symétrisé	slecht afgeregeld
I 4	image-converter	Ultraschallbildwandler *m*	convertisseur *m* d'image	beeldomvormer

I 5	image frequency	Spiegelfrequenz f	fréquence f image	spiegelfrequentie
I 6	image impedance	Ersatzimpedanz f eines Mehrtores	impédance f équivalente	spiegelbeeldimpedantie
I 7	image phase change coefficient	Vierpolwinkelmaß n	coefficient m de glissement de phase	spiegelbeeldfase-exponent
I 8	image phase factor	Vierpolphasenfaktor m	facteur m de phase d'un quadripôle	spiegelbeeldfasefactor
I 9	image response	Spiegelwiedergabe f, Rückwärtsfrequenzgang m	réponse f	spiegelresponsie
I 10	imitation	Nachbildung f	imitation f	imitatie
I 11	immobility characteristics	Unbeweglichkeitskennwerte mpl	caractéristiques fpl d'immobilité	karakteristiek in rust
I 12	immune to vibration	unempfindlich gegen Erschütterung	non sensible aux vibrations	ongevoelig voor trillingen
I 13	impact generator, tapping machine	Hammerwerk n <für Trittschallmessung>	générateur m à marteau, générateur de bruit de frappe	hamer-apparaat n
I 14	impact protection margin	Trittschallschutzmaß n	marge f de protection contre les bruits de pas	speelruimte voor contactgeluidisolatie
I 15	impact sound, footfall sound	Trittschall m	bruit m de pas	contactgeluid n
I 16	impact-sound level	Trittschallpegel m	niveau m de bruit de pas	contactgeluidpeil n
I 17	impact-sound reducing material	trittschalldämpfendes Material n	matériel m absorbant le bruit de pas	isolatiemateriaal n tegen contactgeluid
I 18	impairment of hearing <for conversational speech>	Gehörbeeinträchtigung f <für Umgangssprache>	dégradation f de l'ouïe pour la parole normale	gehoorverzwakking <voor gesprekken>
I 19	imposed oscillation	aufgedrückte (aufgeprägte) Schwingung f	oscillation f imposée	opgedrukte trilling
I 20	impulse noise	Impulsgeräusch n, Impulsrauschen n	bruit m d'impulsion	impulsgeruis n
I 21	impulse period	Impulsperiode f, Pulsfrequenzdauer f	période f d'impulsion	impulsperiode
I 22	impulse repeater	Impulsübertrager m, Impulswiederholer m	répéteur m d'impulsion	impulsenherhaler
I 23	impulse train	Impulsreihe f, Puls m	train m d'impulsions	impulsentrein
I 24	impulsive sound	impulsiver Schall m, impulsives Geräusch n	son m pulsatoire	pulserend geluid n
I 25	inaccurate tuning	ungenaue (unsaubere) Abstimmung f	accord m imparfait (inexact)	onzuivere afstemming
I 26	inadvertent detuning	zufällige Verstimmung f	désaccord m accidentel	ongewilde verstemming
I 27	inarticulate	undeutlich	inarticulé	ongearticuleerd
I 28	inaudibility	Unhörbarkeit f	inaudibilité f	onhoorbaarheid
I 29	incide	einfallen <Strahl>	incider <rayon>	invallen
I 30	incidence	Einfall m s. A 235	incidence f	inval
I 31	incidence angle incidental music	Begleitmusik f	musique f d'accompagnement (de fond sonore)	begeleidende muziek
I 32	incoming frequency	Empfangsfrequenz f, Eingangsfrequenz f	fréquence f d'entrée	ingangsfrequentie
I 33	incoming signal, input signal	Empfangssignal n, Eingangssignal n	signal m d'entrée	ingangssignaal n
I 34	incoming-signal level	Empfangssignalpegel m	niveau m de signal d'entrée	peil n van het ingangssignaal
I 35	incoming wave	einfallende (ankommende) Welle f	onde f d'arrivée	binnenkomende golf
I 36	increasing-value curve	Anstiegskurve f, Flankenanstieg m	flanc m montant (d'une courbe)	oplopende kromme
I 37	indirect echo	Falschecho n	écho m indirect	indirecte echo
I 38	indistinct selectivity	unscharfe Trennung f	sélectivité f peu aiguë	onscherpe selectiviteit
I 39	induced noise	induziertes Geräusch n	bruit m induit	geïnduceerd geruis n
I 40	inductance unbalance	Induktivitätsunsymmetrie f	inductivité f asymétrique	asymmetrie in de zelfinducties
I 41	induction loudspeaker, inductor loudspeaker	Freischwinger m, elektromagnetischer Lautsprecher m	haut-parleur m électromagnétique	elektromagnetische luidspreker
I 42	inductive sound production	induktive Schallerzeugung f	production f de sons induits	inductieve geluidproduktie
	inductor loudspeaker	s. I 41		
I 43	industrial noise	Lärm m in Werkhallen	bruits mpl industriels	fabriekslawaai n
I 44	inertia resistance	Trägheitswiderstand m $\ulcorner f$	résistance f d'inertie	traagheidsweerstand
I 45	infinite attenuation	unendlich große Dämpfung	atténuation f infinie	oneindig grote verzwakking
I 46	inflection infrasonics	Modulation f <Musik> s. I 48	modulation f	stembuiging
I 47	infrasonic wave	Infraschallwelle f	onde f infra-sonique	infrasonore golf
I 48	infrasound, infrasonics inharmonious	Infraschall m s. D 131	infra-son m	infrageluid n
I 49	inherent distortion	Eigenverzerrung f	distorsion f propre	eigenvervorming
I 50	inherent noise	Eigenrauschen n	bruit m propre	eigenruis
I 51	inherent noise pressure	Eigenrauschschalldruck m	pression f sonore de bruit propre	eigengeruisdruk
I 52	initial consonant articulation	Anlautverständlichkeit f	articulation f initiale des consonnes	aanzet-articulatie
I 53	inphase amplifier	Gleichtaktverstärker m	amplificateur m en phase	versterker in gelijke fase
I 54	inphase recording	Gleichtaktaufnahme f	enregistrement m en phase	registratie in gelijke fase
I 55	inphase rejection	Unterdrückung f gleichphasiger Signale	élimination f des signaux de même phase	onderdrukking van de gelijke fase
I 56	inphase signal	Gleichtaktsignal n	signal m en phase	signaal n in gelijke fase
I 57	input amplifier	Vorverstärker m, Eingangsverstärker m	préamplificateur m	voorversterker
I 58	input circuit	Eingangskreis m, Eingangsschaltung f	circuit m d'entrée	ingangskring
I 59	input noise	Eingangsrauschen n	bruit m à l'entrée	ruis aan de ingang van de versterker

I 60	input resonator	Eingangsresonator m	résonateur m d'entrée	ingangsresonator
I 61	input selector	Eingangswählschalter m	sélecteur m d'entrée	keuzeschakelaar aan de ingang
I 62	input signal	s. I 33		
I 62	insert capsule	Einsatzkapsel f	capsule f insérable	insteekkapsel n
I 63	insert cartridge	Eingangswandler m <beim Tonabnehmer>	senseur m en cartouche	insteekelement n
I 64	insert earphone, insert receiver	Einsteckhörer m	écouteur m fichable	insteektelefoon
I 65	insertion loss	Einfügedämpfung f	charge f insérable	tussenschakeldemping
	insert receiver	s. I 64		
I 66	insert transmitter	Kapselmikrofon n	microphone m à capsule	kapselmicrofoon
I 67	inside spider	Zentriervorrichtung f <bei elektrodynamischen Lautsprechern>	spider m, araignée f de centrage	centreerspin
I 68	insonorous material	schalltoter Werkstoff m	matériel m insonore	geluiddempend materiaal n
I 69	instantaneous acoustical speech power	Augenblickswert m der Sprechleistung	puissance f acoustique modulée instantanée	momentaan spraakvermogen n
I 70	instantaneous acoustic kinetic energy	momentane kinetische Schallenergiedichte f	énergie f acoustique cinétique instantanée	momentane akoestische kinetische energie
I 71	instantaneous acoustic power	momentane Schalleistung f	puissance f acoustique instantanée	momentaan akoestisch vermogen n
I 72	instantaneous particle displacement	momentaner Schallausschlag m, momentane Partikelauslenkung f	déplacement m momentané de particules	momentane deeltjesverplaatsing
I 73	instantaneous power output	momentane Ausgangsleistung f	puissance f de sortie instantanée	momentane uitgangsvermogen n
I 74	instantaneous sound energy density	momentane Schallenergiedichte f	densité f instantanée d'energie sonore	momentane geluidenergiedichtheid
I 75	instantaneous sound particle velocity	momentane Schallschnelle f	vélocité f instantanée du son	momentane deeltjessnelheid
I 76	instantaneous sound pressure	momentaner Schalldruck m	pression f sonore instantanée	momentane geluiddruk
I 77	instantaneous speech power	momentane Sprachleistung f	puissance f modulée instantanée	momentane spraakvermogen n
I 78	instrument	instrumentieren <Musik>	instrumenter	instrument n
I 79	insufficient modulation	Untersteuerung f	sousmodulation f	onvoldoende modulatie
I 80	insulation	Dämmung f, Schalldämmung f	isolement m acoustique	isolatie
I 81	insulation curve	Dämmkurve f	courbe f d'atténuation	isolatiekarakteristiek
I 82	intelligence	Signalgehalt m, Modulationsinhalt m	information f	informatie
I 83	intelligence signal	Nutzsignal n	signal m utile	signaal n met informatie
	intelligibility	s. A 237		
I 84	intelligibility of phrases, phrase intelligibility, articulation of sentences	Satzverständlichkeit f	intelligibilité f de phrase	verstaanbaarheid van zinnen
I 85	intelligibility of syllables	Silbenverständlichkeit f	intelligibilité f de syllabes	verstaanbaarheid van lettergrepen
	intelligibility of words	s. A 238		
I 86	intelligible	verständlich	intelligible	verstaanbaar
I 87	intensity level, IL	Intensitätspegel m	niveau m d'intensité	intensiteitspeil n
I 88	intensity of noise, noise intensity	Geräuschstärke f	intensité f de bruit	ruisintensiteit
I 89	interaction crosstalk coupling	Gesamtnebensprechkopplung f	diaphonie f totale de couplage	koppelingsoverspraak
I 90	interaural phase difference	Phasendifferenz f zwischen beiden Ohren <Zeitdifferenz>	différence f de phase interaurale	interauraal faseverschil n
I 91	intercarrier interference	Differenzträgerstörung f	interférence f de signal porteur	interferentie tussen draaggolven
I 92	intercarrier noise suppression	Rauschunterdrückung f	suppression f du bruit de porteur	onderdrukking van ruis tussen de draaggolven
I 93	intercept	abhören	intercepter, capter	onderscheppen
I 94	interception noise, partition noise	Stromverteilungsrauschen n	bruit m de distribution de courant, répartition f du bruit	verdelingsruis
I 95	interchannel spacing	Kanalabstand m	espacement m de canaux	onderlinge kanaalafstand
I 96	interfere	stören	interférer	storen, interfereren
I 97	interference	Störung f	interférence f	interferentie, storing
I 98	interference area	Störgebiet n	zone f d'interférence	interferentiegebied n
I 99	interference elimination	Entstörung f	suppression f d'interférence	storingsonderdrukking
I 100	interference eliminator	Störschutz m	éliminateur m d'interférence	storingsonderdrukker
I 101	interference field, disturbing field	Störfeld n, Interferenzfeld n	champ m d'interférence	storingsgebied n
I 102	interference-free	störfrei, frei von Interferenzen	libre d'interférence	storingsvrij, vrij van interferenties
I 103	interference microphone	Rücksprechmikrofon n	microphone m de retour	commando-microfoon
I 104	interference pattern	Interferenzbild n	diagramme m d'interférence	interferentiepatroon n
I 105	interference range	Störbereich m	plage f d'interférence	stoorafstand
I 106	interference wave, disturbing wave	Störwelle f	onde f interférente (perturbante)	storende golf
I 107	interfering mode	Störmode f	mode m interfératoire	storende trillingswijze
I 108	interfering sound	Störton m	son m interférent	bijgeluid n
I 109	interferometer path	Interferometerstrecke f	ligne f interférométrique	interferometerpad n
I 110	interleave	verschachteln	intercaler	tussenvoegen
I 111	interleaving	Verschachtelung f	intercalation f	tussenvoeging
I 112	interleaving effect	Verschachtelungseffekt m	effet m intercalaire	inlaseffect n
I 113	intermediate echo suppressor	Zwischenechosperre f	suppresseur m d'écho intermédiaire	echo-onderdrukker halverwege
I 114	intermediate frequency	Zwischenfrequenz f	fréquence f intermédiaire, moyenne fréquence	middenfrequentie
I 115	intermediate repeater	Zwischenverstärker m	amplificateur m intermédiaire (de ligne)	tussenversterker

I 116	intermediate waves	Grenzwellen *fpl*	ondes *fpl* intermédiaires	middengolven *pl*
I 117	intermodulation	Zwischenmodulation *f*	intermodulation *f*	kruismodulatie
I 118	intermodulation distortion	Kreuzmodulation *f*	distorsion *f* d'intermodulation	vervorming door kruismodulatie
I 119	intermodulation noise	Klirrgeräusch *n*	bruits *mpl* parasitaires d'intermodulation	ruis door kruismodulatie
I 120	internal losses	Eigenverluste *mpl*	pertes *fpl* internes	inwendige verliezen *npl*
I 121	interquartile range	wahrscheinliche Schwankung *f* <Lautstärkebewertung>	variation *f* probable	waarschijnlijke variaties *pl*
I 122	inter-record gap	Aufzeichnungslücke *f*, Satzlücke *f*	blanc *m* d'enregistrement	registratiegat *n*
I 123	interrupted wave	abgeschnittene Welle *f*, zerhackte Welle	onde *f* interrompue	onderbroken golf
I 124	interruption frequency	Unterbrechungsfrequenz *f*	fréquence *f* d'interruption	onderbrekingsfrequentie
I 125	interruption noise	Unterbrechungsgeräusch *n*	bruit *m* d'interruption	onderbrekingsruis
I 126	interstage	Zwischenstufe *f*	marche *f* (étage *m*) intermédiaire	tussentrap
I 127	interstation noise muting control	Stummabstimmregler *m*	régleur *m* silencieux d'accord	stille afstemming
I 128	intertone	Mischton *m*, Zwischenton *m*	interton *m*	mengtoon
I 129	intertrack bond time displacement	Gegenläufigkeit *f* von Signalen verschiedener Spuren	glissement *m* à contresens des signaux de pistes différentes	tijdverschil *n* tussen de sporen
I 130	interval	Intervall *n*, Pause *f*	intervalle *m*, pause *f*	interval *n*, pauze
I 131	interval signal	Pausenzeichen *n*	indicatif *m*	pauze-teken *n*
I 132	intone	unbetont, monoton	monotone	monotoon
I 133	intrinsic impedance	Kennimpedanz *f*, Wellenwiderstand *m* des freien Raumes, Feldimpedanz *f*	impédance *f* de champ	eigen impedantie
I 134	intrusiveness inverse feedback	Belästigung *f* s. C 203	intrusion *f*	opdringerigheid
I 135	inverse feedback amplifier	Gegenkopplungsverstärker *m*	amplificateur *m* à contreréaction (réaction négative)	tegenkoppelingsversterker
I 136	inversion of common chord	Umkehrung *f* eines Dreiklangs	inversion *f* d'accord	omkering van een drieklank
I 137	inverted crosstalk	unverständliches Nebensprechen *n*	intermodulation *f* inversée	omgekeerde overspraak
I 138	inverted mordent	Pralltriller *m* <Musik>	mordant *m* inversé	praltriller
I 139	inverted speech	invertierte Sprache *f*	parole *f* inversée	omgekeerde spraak
I 140	ionic loudspeaker	Ionenlautsprecher *m*	haut-parleur *m* ionique	ionenluidspreker
I 141	ionic microphone	Ionenmikrofon *n*	microphone *m* ionique	ionenmicrofoon
I 142	ionophone	Ionofon *n*, Plasmastrahler	ionophone *m*	ionofoon
I 143	ionospheric sounding	Ionosphärenlotung *f*	sondage *m* de l'ionosphère	echopeiling in de ionosfeer
I 144	irradiate acoustically	Schall einstrahlen, beschallen	irradier acoustiquement, sonoriser	akoestisch bestralen
I 145	irradiated sound wave	eingestrahlte Schallwelle *f*	onde *f* sonore irradiée	bestralende geluidgolf
I 146	irrotational	wirbelfrei	non tourbillonnaire	irrotationeel, zonder wervels
I 147	irrotational field	wirbelfreies Feld *n*	champ *m* non tourbillonnaire	irrotationeel veld *n*
I 148	isolation	Isolierung *f*	isolement *m*	isolatie
I 149	isophonic contour issuing ray	Isofoniekurve *f* s. E 56	courbe *f* isophonique	isofoon
I 150	iterative attenuation	Kettendämpfung *f*	atténuation *f* en chaîne	geïtereerde verzwakking
I 151	iterative attenuation factor	Kettendämpfungsfaktor *m*	facteur *m* d'atténuation en chaîne	geïtereerde-verzwakkingscoëfficiënt
I 152	iterative impedance	Kettenimpedanz *f*, Wellenwiderstand *m* <eines Vierpols>	impédance *f* de chaîne	geïtereerde impedantie
I 153	iterative transfer constant	Kettenübertragungsmaß *n*	constante *f* itérative de transfert, exposant *m* itératif de propagation	geïtereerde-doorlaatcoëfficiënt

J

J 1	jack	Docke *f* <Musik>, Klinke *f* <Stecker>	sauteau *m* <cembalo>, jack *m*, fiche *f*, cheville *f*	dokje *n* <clavecimbel>, plug <telefoon>
J 2	jam	stören	troubler, presser, coincer	storen
J 3	jammer	Störer *m*, Störsender *m*	gêneur *m*, troubleur *m*	storingsbron
J 4	jamming	[beabsichtigte] Störung *f*	troublage *m*	verstoring
J 5	jerk	Ruck *m*	saccade *f*, secousse *f*	schok
J 6	jet-edge generator	Düsengenerator *m*	générateur *m* à buse	fluittoongenerator
J 7	jew's harp	Maultrommel *f*, Brummeisen *n*	guimbarde *f*	mondtrom
J 8	jingles	Schellen *fpl* <Musik>	grelots *mpl*	schellen
	jingling Johnny <sl>	s. B 105		
	jnd.	s. J 12		
	Johnson noise	s. T 33		
J 9	joint	Cutterstelle *f* <Tonband>	joint *m*	las
J 10	juke-box	Musikbox *f*	juke-box *f*	muziekautomaat
J 11	junctional tissue	Reizleitungssystem *n*	chaîne *f* d'excitation sensorielle	zenuwweefsel *n*
J 12	just noticeable difference, jnd.	Wahrnehmbarkeitsstufe *f*	seuil *m* de perceptibilité	trede
J 13	just scale	reine Stimmung *f*	gamme *f* pure	goed getempereerde toonladder
	just scale	s. a. M 28		

K

K 1	kazoo	Summrohr n <Rohr mit Darmsaite>	kazoo m <mirliton à corde>	mirleton
K 2	keep time	Takt halten	conserver la mesure	maat houden
K 3	kettledrum	Kesselpauke f	timbale f	pauk
K 4	key	Klappe f, Taste f <Musik>	clé f	klep, sleutel, toets
K 5	keyboard, bank of keys	Klaviatur f, Tastatur f	clavier m, claviature f	toetsenbord n
K 6	key bugle	Klapphorn n	bugle m	pistontrompet
K 7	key chamber	Tonkanzelle f <Orgel>	laye f	windlade <orgel>
K 8	key signature	Vorzeichnung f <Musik>	signe m de clé	sleutelvoortekening
	kid <sl>	s P 136		
	krummhorn	s. C 227		
K 9	Kundt's tube	Kundtsches Rohr n	tube m de Kundt	buis van Kundt

L

L 1	labial pipe, flue pipe	Labialpfeife f, Lippenpfeife f <Orgel>	tuyau m d'orgue labial, flûte f douce	labiaalpijp
L 2	labial sound	Lippenlaut m	consonne f labiale, labiale f, son m labial	labiaalklank
L 3	labium	Lippe f <Orgelpfeife>	anche f	tong <orgelpijp>
L 4	labyrinth loudspeaker	Lautsprecher m mit akustischem Labyrinth	haut-parleur m à labyrinthe	labyrintluidspreker
	labyrinth of loudspeaker	s. D 42		
L 5	lack of definition	Unschärfe f	manque m de netteté	onscherpte
L 6	lack of depth	Tiefenunschärfe f	manque m de netteté en profondeur	gebrek n aan diepte
	lagging of phase	s. P 83		
L 7	lapel microphone	Knopflochmikrofon n	microphone m de boutonnière	knoopsgatmicrofoon
L 8	laryngeal ligament	Kehlkopfband n	tendon m laryngual	strottehoofdband
L 9	laryngograph	Laryngograf m	laryngographe m	laryngograaf
	laryngophone	s. N 15		
L 10	larynx	Kehlkopf m	larynx m	strottehoofd n
L 11	lateral recording	seitliche Aufzeichnung f, Seitenschrift f	enregistrement m latéral	laterale opname
L 12	lattice filter	Kreuzgliedfilter n	cellule f filtrante en treillis	kruisnetwerkfilter
L 13	lattice network, bridge network	Vierpolkreuzglied n	cellule f en treillis [quadripôle]	brugschakeling
L 14	law of chance	Zufallsgesetz n	loi f du hasard	toevalswet
L 15	leader tape	Vorspannband n	ruban m initial	beginstrook
L 16	leading tone	Leitton m	ton m conducteur	leitoon
L 17	lead screw	Leitspindel f <Schneidedose>	arbre m fileté, vis f de chariot	transportschroef
L 18/9	leakage attenuation	Ableitungsdämpfung f	atténuation f par pertes, amortissement m de perditance	lekverliezen npl
L 20	leakage coupling	Streukopplung f	couplage m de dispersion	koppeling door verstrooiing
L 21	leakage field	Streufeld n	champ m de dispersion	strooiveld n
L 22	leakage impedance	Streuimpedanz f	impédance f de dispersion	verstrooiingsimpedantie
L 23	leak impedance	Querimpedanz f, Ableitungsimpedanz f	impédance f de fuite	lekimpedantie
L 24	leak resistance	Ableitungsresistanz f	résistance f de fuite	lekweerstand
L 25	left-handed polarization	Linkspolarisation f	polarisation f lévogyre	linksdraaiende polarisatie
	left-handed polarized wave	s. C 202		
L 26	left hand volume	Lautstärke f linker Kanal <Stereo>	puissance f sonore du canal gauche	volume n van het linker kanaal
L 27	level	Pegel m	niveau m	peil n, niveau n
L 28	level difference	Schallpegeldifferenz f	différence f de niveau	peilverschil n, niveauverschil n
L 29	level meter	Pegelmesser m	niveaumètre m <de transmission>	peilmeter, niveaumeter
L 30	level recorder	Pegelschreiber m	enregistreur m de niveau	niveauschrijver
L 31	light-localising pick-up	elektrooptischer Tonabnehmer m	senseur m électro-optique	elektro-optische groeftaster
L 32	light-valve	Lichtmodulator m	modulateur m de lumière	lichtmodulator
L 33	limen	Schwellwert m	seuil m	trede
L 34	limitation of frequency band	Frequenzbandbegrenzung f	limitation f de bande de fréquences	begrenzing van de frequentieband
L 35	limitation of interference	Störbegrenzung f	limitation f d'interférence	storingsbegrenzing
	limit frequency	s. C 245		
L 36	limiting amplifier	Begrenzungsverstärker m, Begrenzer m	amplificateur m limiteur	begrenzende versterker
L 37	limiting ray	Grenzstrahl m	rayon m limite	begrenzende straal
L 38	limiting sensitivity	Grenzempfindlichkeit f	sensibilité f limite	grensgevoeligheid
L 39	limit of audibility	Hörbarkeitsgrenze f	limite f d'audibilité	hoorbaarheidsgrens
L 40	limit of definition	Schärfefeldgrenze f	limite f de définition	grens van het oplossend vermogen
L 41	linear amplification	lineare Verstärkung f	amplification f linéaire	lineaire versterking
L 42	linear distortion	lineare Verzerrung f	distorsion f linéaire	lineaire vervorming
L 43	linear system <transducer>	lineares System n <Übertrager>	système m linéaire <transducteur>	lineair systeem n <transducent>
L 44	line attenuation	Leitungsdämpfung f	amortissement m de ligne	lijndemping, leidingsdemping
L 45	line hydrophone	Linienhydrofon n, linienförmiges Hydrofon n	hydrophone m de ligne	lijnvormige hydrofoon
L 46	line microphone	Linienmikrofon n, lineares Mikrofon n	groupe m linéaire de microphones	lijnvormige microfoon
L 47	line resistance	Leitungsresistanz f	résistance f de ligne	leidingsweerstand

L 48	line spectrum	Linienspektrum n	spectre m de lignes	lijnenspectrum n
L 49	line trap	Wellensperre f	ligne f de réjection, piège m à ondes	zeefkring
L 50	line up	einpegeln	régler le niveau	uitlijnen
L 51	lingering sound	Nachklang m, Nachhall m	écho m	nagalm
L 52	lip microphone	Lippenmikrofon n	microphone m labial	lipmicrofoon
L 53	lip reading, visual reading	Ablesen n von den Lippen	lecture f sur les lèvres	liplezen
L 54	liquid vortex tones	Töne mpl, die durch Flüssigkeitswirbel erzeugt sind	sons mpl produits par tourbillons liquides	tonen pl ontstaan door vloeistofwervels, draaikolktonen pl
L 55	listen	horchen, lauschen, zuhören	écouter, épier	luisteren
	listener-in	s. B 185		
L 56	listen in	Radio hören, zuhören, hineinhören	écouter	uitluisteren
L 57	listening area	Hörgebiet n	aire f d'audition	luistergebied n
L 58	listening conditions	Hörbedingungen fpl	conditions fpl d'écoute	luistercondities pl
L 59	listening position	Hörort m	point m d'écoute	luisterpositie
L 60	listening sonar	passiver Schallempfang m ‹aktive Schallquelle›	réception f passive des ondes d'un sonar	passief luisterende sonar
L 61	live end	Reflexionswand f	section f terminale réfléchissante	reflecterende wand
L 62	liveness	Halligkeit f	résonance f	hardklinkendheid
L 63	liven the studio	Schallreflexionen erzeugen ‹im Studio›	produire de la résonance en studio	verhoging van het echoeffect van een studio
L 64	live room	halliger Raum m ‹Wohnraum›	chambre f à écho	harde kamer
L 65	live studio	halliges Studio n	studio m à écho	studio met hoog echo-effect
L 66	loaded impedance	Impedanz f bei Sollabschluß	impédance f adaptée	ingangsimpedantie bij nominale belasting
L 67	load impedance	Außenwiderstand m, Belastungsimpedanz f	impédance f de charge	belastingsimpedantie
	lobe	s. R 10		
L 68	local fading, near fading	Nahschwund m	fading m local (de proximité)	lokale fading
L 69	localization	Ortung f	localisation f	plaatsbepaling
L 70	local oscillation	Überlagerungsschwingung f	oscillation f locale	mengoscillatie, mengtrilling
L 71	local oscillator	Überlagerungsoszillator m	oscillateur m local	mengoscillator
L 72	local-oscillator frequency	Überlagererfrequenz f	fréquence f locale (de superposition, d'hétérodyne)	mengfrequentie
L 73	locked synchronization	Synchronisierzwang m	contrainte f de synchronisation	vergrendelde synchronisatie
L 74	lock-in amplifier	Einfangverstärker m, synchronisierter Verstärker m	amplificateur m de synchronisation	synchrone versterker
L 75	locking tone	Synchronisierton m	ton m de synchronisation	synchronisatietoon
L 76	locking voltage	Mitnahmespannung f	tension f de synchronisation	synchronisatiespanning
L 77	logarithmic amplifier	logarithmischer Verstärker m	amplificateur m logarithmique	logarithmische versterker
L 78	logarithmic decrement	logarithmisches Dekrement n	décrément m logarithmique	logarithmisch decrement n
L 79	logarithmic horn	Exponentialrichter m ‹Lautsprecher› ⌐m	pavillon m exponentiel	exponentiële hoorn
L 80	logarithmic receiver	logarithmischer Empfänger m	récepteur m logarithmique	logarithmische ontvanger
L 81	logatom	Logatom n	logatom m	logatoom n
L 82	longitudinal wave	Longitudinalwelle f	onde f longitudinale	longitudinale golf
L 83	loop, antinode	Schwingungsbauch m, Bauch m	ventre m	buik
L 84	loop gain	Rückkopplungsschleifenverstärkung f	gain m de boucle de réaction	rondgaande versterking
L 85	loose coupling	lose Kopplung f	couplage m mou (faible)	losse koppeling
L 86	loosely coupled resonant line	lose angekoppelte Resonanzleitung f	lignes fpl de résonance faiblement couplées	los gekoppelde afgestemde leiding
L 87	loss factor	Verlustfaktor m	facteur m de pertes	verliesfactor
L 88	loss of sound	Tonausfall m	blanc m	wegvallen van het geluid
L 89	lossy	verlustbehaftet	affecté de pertes	met inwendige verliezen
L 90	loudness	Lautheit f	puissance f de son, sonorité f	luidheid
L 91	loudness contour of ear	Ohrempfindungsgrenze f, Schwellwertkurve f ‹des Ohres›	courbe f de sensibilité de l'oreille	isofoon
L 92	loudness judgment	Lautheitsbeurteilung f	appréciation f d'intensité sonore	beoordeling van de luidheid
L 93	loudness level	Lautstärkepegel m	niveau m sonore, niveau d'intensité sonore	luidheidspeil n
L 94	loudness-level contour	Kurve f gleichen Lautstärkepegels	courbe f d'intensité sonore constante	isofoon
L 95	loudness sensation	Lautheitsempfindung f	sensation f d'intensité sonore	waarneming van luidheid
L 96	loudness summation	Lautheitssummation f	sommation f de niveau sonore	luidheidsommering
L 97/8	loudness volume equivalent	äquivalente Lautheit f	intensité f sonore équivalente	equivalente luidheid
	loud pedal	s. D 7		
L 99	loudspeaker, speaker ‹US›	Lautsprecher m	haut-parleur m	luidspreker
L 100	loudspeaker cloth	Lautsprecherbespannung f	tenture f (étoffe f) de haut-parleur	luidsprekerdoek n
L 101	loudspeaker diaphragm	Lautsprechermembran f	membrane f de haut-parleur	luidsprekermembraan n
L 102	loudspeaker dividing network	Frequenzweiche f ‹des Lautsprechers›	filtre m d'aiguillage pour haut-parleur	overgangsfilter n voor luidsprekers
L 103	loudspeaker horn	Lautsprechertrichter m	pavillon m de haut-parleur	luidsprekerhoorn
L 104	loudspeaker van	Lautsprecherwagen m	fourgon m pour haut-parleurs	luidsprekerwagen
L 105	loudspeaker voice coil, voice coil	Schwingspule f ‹des Lautsprechers›	bobine f mobile [de haut-parleur]	spreekspoel

L 106	**loudspeaker with tweeter dome**	Lautsprecher *m* mit Hochtonkegel	haut-parleur *m* à diffuseur d'aiguës	luidspreker met discantkapje
L 107	**loudspeaking telephone**	Lautfernsprecher *m*	téléphone *m* haut-parleur	luidsprekende telefoon
L 108	**loudspeaking telephone without voice**	Lautfernsprecher *m* ohne Sprachsteuerung	téléphone *m* haut-parleur sans commande vocale	luidsprekende telefoon zonder stemcommando
L 109	**louvre**	Schallöffnung *f* <Lautsprecher>	orifice *m*	luidspreker-opening
L 110	**low**	tief	bas	laag
L 111	**low-end cut-off frequency**	untere Grenzfrequenz *f*	limite *f* inférieure de bande de fréquences	laagste afsnijfrequentie
L 112	**lower band**	frequenzniedrigeres Band *n*	bande *f* de fréquences plus basse	laagfrequente band
L 113	**lower pitch limit**	untere Frequenzhörgrenze *f*	limite *f* inférieure des fréquences audibles	laagste te onderscheiden toonhoogte
L 114	**lowest useful frequency**	tiefste Nutzfrequenz *f*	plus basse fréquence *f* utilisable	laagste bruikbare frequentie
L 115	**low fence <US>**	Anfangswert *m* der Schwerhörigkeit	limite *f* inférieure de la surdité	laagste hardhorendheidsdrempel
L 116	**low-frequency cone loudspeaker**	Tieftonkonus *m*	haut-parleur *m* à cône pour sons graves	conusluidspreker voor lage tonen
	low-frequency loudspeaker	*s.* B 159		
L 117	**low-frequency muffler**	Tieffrequenzsperre *f*	suppresseur *m* de sons graves	geluiddemper voor lage frequenties
L 118	**low-frequency suppression filter**	Tieftonunterdrücker *m*	filtre *m* suppresseur de basses fréquences	afsnijfilter *n* voor lage tonen
L 119	**low-level amplifier**	Grundpegelverstärker *m*	amplificateur *m* de niveau de base	versterker voor zwakke signalen
L 120	**low-level blanking**	Grundgeräuschaustastung *f*	suppresseur *m* de bruit de fond	onderdrukker voor achtergrondruis
L 121	**low-loss**	verlustarm	à pertes réduites	verliesarm
L 122	**low-pass filter**	Tiefpaßfilter *n*	filtre *m* passe-bas	laagdoorlaatfilter *n*
L 123/4	**low-pitched tone**	tiefer Ton *m*	son *m* grave	lage toon
L 125	**low-resistance**	niederohmig	à résistance basse	laagohmig
L 126	**lumped-parameter system**, discrete-parameter system	System *n* mit endlicher Anzahl von Freiheitsgraden	système *m* à nombre limité (défini) de degrés de liberté, système à paramètres discrets	systeem *n* met beperkt aantal vrijheidsgraden
L 127	**lute**	Laute *f* <Musik>	luth *m*	luit

M

M 1	**macrosonics**, non-linear acoustics	nichtlineare Akustik *f*, Akustik *f* hoher Pegel	acoustique *f* non linéaire	niet-lineaire akoestiek
M 2	**magnetic biasing**	Vormagnetisierung *f*	pré-aimantation *f*	voormagnetisering
M 3	**magnetic clamp**	Haftmagnet *m*	aimant *m* de retenue	magnetische klem
M 4	**magnetic head alignment**	Magnetkopfjustierung *f*	alignement *m* de tête magnétique	justeren van de magneetkoppen
M 5	**magnetic hum**	magnetischer Brumm *m*	ronflement *m* magnétique	magnetische brom
M 6	**magnetic loudspeaker**	magnetischer Lautsprecher *m* ⌐*m*	haut-parleur *m* magnétique	magnetische luidspreker
M 7	**magnetic pick-up**	magnetischer Tonabnehmer *m*	lecteur *m* magnétique	magnetische opnemer
M 8	**magnetic-plated wire**	magnetischer (mit magnetischem Material beschichteter) Draht *m*	fil *m* à couche magnétique	magnetisch-geplatteerde draad
M 9	**magnetic recorder**	Magnetaufzeichnungsgerät *n*	enregistreur *m* magnétique	magnetische recorder
M 10	**magnetic recording**	magnetische Tonaufzeichnung *f*	enregistrement *m* magnétique	magnetische opname
M 11	**magnetic sheet**	Magnettonplatte *f*	disque *m* magnétique	magnetische schijf
M 12	**magnetic sound**	Magnetton *m*	son *m* magnétique	magnetisch geluid *n*
M 13	**magnetic sound recorder**	Magnettongerät *n*	enregistreur *m* de sons magnétiques	magnetische geluidopnemer
M 14	**magnetic tape cross talk**	Kopiereffekt *m*	copiage *m* par superposition	doordrukeffect *n*, echoeffect *n*
M 15	**magnetic tape recorder**, tape recorder	Tonbandgerät *n*	enregistreur *m* à bande magnétique	magnetische-bandrecorder
M 16	**magnetic wire**	Magnetdraht *m* <für Schallaufzeichnung>	fil *m* magnétique	magnetische draad
M 17	**magnetostriction loudspeaker**	Magnetostriktionsschallstrahler *m*	haut-parleur *m* magnétostrictif	magnetostrictieve luidspreker
M 18	**magnetostriction microphone**	magnetostriktives Mikrofon *n*	microphone *m* magnétostrictif	magnetostrictieve microfoon
M 19	**magnetostriction vibrator**	Magnetostriktionsschwinger *m*	vibrateur *m* magnétostrictif	magnetostrictieve triller
M 20	**magnetostrictive probe microphone**	magnetostriktives Sondenmikrofon *n*	microphone *m* magnétostrictif de sondage	magnetostrictieve sondemicrofoon
M 21	**magnetostrictive transceiver**	magnetostriktiver Empfänger *m* und Sender *m*	transceiver *m* magnétostrictif	magnetostrictieve opnemer en weergever
M 22	**main lobe**	Hauptkeule *f*	lobe *m* principal	hoofdbundel
M 23	**mains hum**	Netzbrumm *m*	ronflement *m* de secteur	netbrom
M 24	**mains noise**	Netzgeräusch *n*	parasites *mpl* de secteur	netruis
M 25	**maintained tuning fork**	Stimmgabelgenerator *m*	diapason *m* à branches	stemvorkgenerator
M 26	**major scale**	Dur-Tonleiter *f*	gamme *f* majeure	majeurtoonladder
M 27	**major scale of equal temperament**	größere gleichtemperierte Tonleiter *f*	grande gamme *f* uniformément tempérée	getempereerde majeurtoonladder
M 28	**major scale of just temperament**, just scale	wohltemperierte Tonleiter *f*	gamme *f* bien tempérée	goed getempereerde toonladder
M 29	**major third**	große Terz *f*	grande tierce *f*	grote terts
M 30	**major triad**	Dur-Dreiklang *m*	accord *m* majeur	majeurdrieklank
M 31	**mandola**	Mandora *f*, Mandola *f*, viersaitige Laute *f*	mandore *f*, mandoline *f* à quatre cordes	mandola

M 32	man-made radio interference	Störung f durch elektrische Geräte	parasitage m par appareils électriques	niet-natuurlijke radiostoring
M 33	manual	Manual n <der Orgel>	manuel m	manuaal n
M 34	manual coupler	Manualkoppler m	enclenchement m de manuel	manuaalkoppeling
M 35	maraca	Rumbakugel f	maraca f	samba-bal
M 36	marginal track	Randspur f	trace f marginale	randspoor n
M 37	mark	Marke f, Kennzeichnung f	marque f	merkteken n
M 38	marked	ausgeprägt, deutlich, markant	marqué, marquant	markant
M 39	marker generator	Frequenzmarkengenerator m	générateur m de fréquence à marques multiples	markeringsgenerator
M 40	mask	überdecken, abdecken, maskieren	masquer	maskeren
M 41	masking	Verdeckung f	masque m	maskering
M 42	masking audiogram	Verdeckungsaudiogramm n	audiogramme m à masque	maskeringsaudiogram n
M 43	masking effect	Verdeckungseffekt m	effet m de masque	maskeringseffect n
M 44	masking level	Verdeckungspegel m	degré m de masquage	maskeerniveau n
M 45	masking sound	verdeckender Schall m	son m masquant (couvrant)	maskerend geluid n
M 46	mask microphone	Maskenmikrofon n	microphone m à masque	maskermicrofoon
M 47	mass-and-spring system	Masse-Feder-System n	système m masse-ressort	massaveersysteem n
M 48	mass impedance	Massenimpedanz f	impédance f de masse	massa-impedantie
M 49	mass reactance	Massenreaktanz f	réactance f de masse	massareactantie
M 50	master	Mater f	maître m, matrice f	matrijs, moeder
M 51	master disk	Vaterplatte f	maître-disque m	vader
M 52	master gain control	Hauptverstärkungsregler m	régleur m général d'amplification	centrale versterkingsregeling
M 53	master studio	Hauptstudio n	studio m principal	centrale studio
M 54	match	anpassen	adapter, accorder, aligner	aanpassen
M 55	matching attenuation	Anpassungsdämpfung f	amortissement m d'adaptation	aanpassingsdemping
M 56	matching transformer	Anpassungsübertrager m	transformateur m d'adaptation	aanpassingstransformator
M 57	matted track	Tonspur f mit einstellbarer Breite	trace f à largeur réglable	geluidsspoor n met veranderlijke dichtheid en breedte
M 58	matter wave	Materiewelle f	onde f matérielle	materiegolf
M 59	mean free path mean frequency	mittlere freie Weglänge f s. C 35	distance f moyenne libre	gemiddelde vrije weglengte
M 60	mean sound pressure level <in a room>	mittlerer Schalldruckpegel m <in einem Raum>	pression f sonore moyenne <dans un espace>	gemiddeld geluiddrukpeil n <in een kamer>
M 61	measure	Takt m <Musik>	mesure f	maat
M 62	mechanical image impedance	mechanische Kennimpedanz f	impédance f image mécanique	mechanische spiegelbeeldimpedantie
M 63	mechanical line	mechanische Leitung f	ligne f mécanique	mechanische transmissielijn
M 64	mechanical resistance	mechanische Resistanz f	résistance f mécanique	mechanische weerstand
M 65	mechanical shock	mechanischer Stoß m	choc m mécanique	mechanische schok
M 66	mechanical vibration	mechanische Schwingung f	vibration f mécanique	mechanische trilling
M 67	mediant	Mediante f	médiante f	mediant
M 68	medium of propagation	Ausbreitungsträger m, Schallmedium n	médium m de propagation	voortplantingsmedium n
M 69	megaphone	Sprachrohr n	mégaphone m	megafoon
M 70	mel <unit>	mel <Einheit>	mel <unité>	mel <eenheid>
M 71	mellifluous	weich <Stimme>	mielleuse <voix>	zoetvloeiend
M 72	melodic	melodisch	mélodique, mélodieux	melodieus
M 73	membrane	Membran f	membrane f	membraan n
M 74	method of adjusting	Angleichmethode f <Lautheitsmessung>	méthode f d'ajustage	methode van afregelen
M 75	method of constant stimulus	Methode f des konstanten Reizes <Lautheitsmessung>	méthode f de l'excitation constante	methode van gelijkblijvende stimulus
M 76	method of tracking	Pendelangleichmethode f <Lautheitsmessung>	méthode f itérative	iteratieve methode
	metronome	s. B 92		
M 77	microgroove	Mikrorille f <Schallplatte>	microsillon m	microgroef
M 78	microphone, mike <sl>	Mikrofon n	microphone m	microfoon
M 79	microphone amplifier, speech-input amplifier	Mikrofonverstärker m	amplificateur m de microphone	microfoonversterker
M 80	microphone blanket	Mikrofonkappe f	cape f de microphone	microfooncape
M 81	microphone boom	Mikrofongalgen m	gibet m de microphone	microfoonhengel
M 82	microphone buzzer	Mikrofonsummer m	buzzer m de microphone	microfoonzoemer
M 83	microphone capsule	Mikrofonkapsel f	capsule f microphonique	microfoonkapsel n
M 84	microphone carbon	Mikrofonkohle f	charbon m de microphone	microfoonkool
M 85	microphone cell	Mikrofonelement n	cellule f microphonique	microfoonelement n
M 86	microphone coupling	Mikrofonkopplung f	accouplement m de microphone	microfoonkoppeling
M 87	microphone hiss	Mikrofonzischen n	chuintement m microphonique	microfoonruis n
M 88	microphone noise	Mikrofongeräusch n	bruit m microphonique	microfoongeruis n
M 89	microphone relay	Mikrofonrelais n	relais m microphonique	microfoonrelais n
M 90	microphone stand	Mikrofonstativ n	support m de microphone	microfoonstatief n
M 91	microphone transformer	Mikrofonübertrager m	transformateur m de microphone	microfoontransformator
	microphone with cardioid characteristic	s. C 15		
M 92	microphonic voltage	Klingspannung f	tension f microphonique	microfonische spanning
M 93	microphony	Mikrofoneffekt m, Mikrofonie f	effet m microphonique (Larsen)	microfonie
M 94	micro therapy	Ultrakurzwellentherapie f	thermothérapie f à ondes très courtes	microgolftherapie
M 95	mid-band	Bandmitte f	milieu m de bande	midden n van de band
M 96	mid-band frequency	Bandmittenfrequenz f	fréquence f médiane	middenfrequentie van de band
M 97	middle voice	Mittelstimme f	voix f moyenne	middenstem
M 98	midget microphone mike	Zwergmikrofon n s. M 78	minimicrophone m	dwergmicrofoon

	English	German	French	Dutch
M 99	mike stew ⟨sl⟩	Mikrofonschmoren n	friture f de microphone	smoren n van de microfoon
M 100	milli-octave	Millioktave f	milli-octave f	millioktaaf n
M 101	minimum audibility	Minimalhörbarkeit f	audibilité f minimum	minimum hoorbaarheid
M 102	minor interval	kleines Intervall n	intervalle m mineur	klein interval n
M 103	minor key	Moll-Tonart f	ton m (mode m) mineur	mineursleutel
M 104	minor scale	Moll-Leiter f, kleine Tonleiter f	gamme f mineure	mineurtoonladder
M 105	misaligned	falsch abgeglichen, schlecht ausgerichtet	désaligné, désaccordé	slecht afgeregeld
M 106	misalignment	Fehlabgleichung f	désalignement m	foutieve afregeling
M 107	miss the beat	aus dem Takt kommen	perdre la mesure	uit de maat raken
	mistune	s. D 65		
	mistuning	s. D 66		
M 108	mix	mischen	mélanger	mengen
M 109	mixed coupling	gemischte Kopplung f	couplage m mixte	gemengde koppeling
M 110	mixture stops	Mixtur f ⟨Orgel⟩	mixture f	mixtuur
M 111	mixture trautonium	Mixtur-Trautonium n	mixture-trautonium m	mixtuurtrautonium n
M 112	modal numbers	Modenzahlen fpl	numéros mpl de modes	ranggetallen npl van trillingswijze
M 113	mode	Schwingungsart f ⟨resonanzfähiger Systeme⟩	mode m de résonance	trillingswijze
M 114	mode of propagation	Ausbreitungsart f	mode m de propagation	voortplantingswijze
M 115	mode of vibration	Schwingungsmode f	mode m de vibration	trillingswijze
M 116	modulate	modulieren, modeln	moduler	moduleren
M 117	modulated wave	modulierte Welle f	onde f modulée	gemoduleerde golf
M 118	modulating frequency	Modelfrequenz f	fréquence f de modulation	modulatiefrequentie
M 119	modulation	Aussteuerung f, Modelung f	modulation f	modulatie
M 120	modulation converter	Modulationswandler m	convertisseur m de modulation	modulatieomvormer
M 121	modulation distortion	Modulationsverzerrung f	distorsion f de modulation	modulatievervorming
M 122	modulation frequency response	Modulationsfrequenzgang m	réponse f en fréquence de modulation	frequentiekarakteristiek van de modulatie
M 123	modulation noise	Modulationsrauschen n	bruit m de modulation	modulatieruis
M 124	modulation suppression	Modulationsunterdrückung f	suppression f de modulation	modulatieonderdrukking
M 125	modulation threshold	Modulationsschwelle f	seuil m de modulation	modulatiedrempel
M 126	modulus of decay	Abklingmodul m	module m de décroissement	uittrillingsmodulus
M 127	molecular relaxation	molekulare Relaxation f	relaxation f moléculaire	moleculaire relaxatie
M 128	molecular resonant absorption	Absorption f durch molekulare Resonanz	absorption f par résonance moléculaire	absorptie door moleculaire resonantie
M 129	moment of force	Kraftmoment n	moment m de force	krachtmoment n
M 130	moment of inertia	Trägheitsmoment n	moment m d'inertie	traagheidsmoment n
M 131	monaural	einohrig	monaural	voor één oor
M 132	monitor	mithören, überwachen	contrôler	controleren
M 133	monitor earphone	Mithörkopfhörer m	écouteur m de monitor	controletelefoon
M 134	monitoring amplifier	Kontrollverstärker m	amplificateur m de contrôle	controleversterker
M 135	monitor loudspeaker, pilot loudspeaker (monitor)	Kontrollautsprecher m, Monitor m	haut-parleur m de contrôle	controleluidspreker
M 136	monophonic	einstimmig, monofon	monophone	monofoon
M 137	monophony	Monofonie f	monophonie f	monofonie
M 138	monopole simple sound source	Punktstrahler m	source f sonore ponctuelle	monopool, puntbron
M 139	monotic	nur ein Ohr betreffend	monoral	voor één oor
M 140	mood	Modus m ⟨Musik⟩	mode m	modus
	mood music	s. B 3		
M 141	mother	Mutterplatte f	disque m mère	moederplaat
M 142	motional impedance	Bewegungsimpedanz f	impédance f de mobilité	bewegingsimpedantie
M 143	motion-picture sound reproduction system	Kinofilm-Tonwiedergabesystem n	système m sonore de reproduction pour filmes cinématographiques	cinema-geluidweergavesysteem n
M 144	motor rumble	Laufgeräusch n (Rumpeln n) des Motors ⟨Antrieb⟩	bruit m de fonctionnement de moteur	motorgestommel n
M 145	mouth	Pfeifenmund m ⟨Orgel⟩	embouchure f	opsnede
M 146	mouth of horn	Trichteröffnung f ⟨Lautsprecher⟩	ouverture f de pavillon	hoornopening
M 147	mouth organ	Mundharmonika f	harmonica f	mondharmonica
M 148	mouthpiece of microphone	Mikrofoneinsprechöffnung f	ouverture f de microphone	inspreekopening van een microfoon
M 149	moving coil	Schwingspule f	bobine f mobile	spreekspoel
M 150	moving-coil headset	elektrodynamischer Kopfhörer m	écouteur m électrodynamique	elektrodynamische hoofdtelefoon
M 151	moving-coil loudspeaker, moving-conductor loudspeaker	elektrodynamischer Lautsprecher m	haut-parleur m électrodynamique (à ruban)	elektrodynamische luidspreker
M 152	moving-coil microphone	Tauschspulenmikrofon n	microphone m à bobine mobile	elektrodynamische microfoon
M 153	moving-coil pick-up	elektrodynamischer Tonabnehmer m	lecteur m électrodynamique	elektrodynamische opnemer
	moving-conductor loudspeaker	s. M 151		
	moving-conductor microphone	s. E 44		
M 154	moving-iron loudspeaker	elektromagnetischer Lautsprecher m	haut-parleur m électromagnétique	elektromagnetische luidspreker
M 155	MS-technique	MS-Schallaufnahmeverfahren n der Stereotechnik	MS-technique f d'enregistrement stéréophonique	MS-techniek
M 156	muffle	bedämpfen, abdämpfen, verstopfen	assourdir, amortir	dempen
M 157	muffled	dumpf, gedämpft	assourdi, amorti	gedempt
M 158	muffler	Schalldämpfer m	sourdine f	demper
M 159	multicellular horn	mehrzelliger Hornlautsprecher m	cellule f de haut-parleurs à pavillon	veelcellige hoorn

	English	German	French	Dutch
M 160	multicellular loud-speaker	Vielzellenlautsprecher m	haut-parleur m multicellu-laire	meercellige luidspreker
M 161	multichannel loud-speaker	Mehrkanallautsprecher m	haut-parleur m milticanaux	luidsprekercombinatie
M 162	multi-delay equipment	Mehrfachverzögerungs-gerät n	équipement m multi-délais	meervoudig vertragingsap-paraat n
M 163	multi-loop feedback amplifier	mehrfach rückgekoppelter Verstärker m	amplificateur m à réaction multiple	meervoudig teruggekoppel-de versterker
M 164	multipath fading	Mehrwegfading n	fading m à voies multiples	multipadfading, fading door multipadverbindingen
M 165	multipath reflection	Mehrwegreflexion f	réflexion f à chemins multiples	multipadreflecties pl
M 166	multiple-degree-of-freedom system	System n mit mehreren Freiheitsgraden	système m à degrés de liberté multiples	systeem n met veel vrijheids-graden
M 167	multiple echo	Mehrfachecho n	écho m multiple	meervoudige echo
M 168	multiple grating	Beugungsgitter n	grille f de réfraction	buigingsrooster n
M 169	multiple series of harmonic sounds	Klanggemisch n, mehr-facher Klang m	sons mpl harmoniques sériels multiples	meervoudige reeks van har-monische geluiden
M 170	multiple sound track	Mehrfachtonspur f	pistes fpl d'enregistrement multiples	meervoudig geluidspoor n
M 171	multiplication amplifier	Vervielfacherverstärker m	amplificateur m de multi-plication	vermenigvuldigingsver-sterker
M 172	multiplicative mixing	multiplikative Mischung f	conversion f multiplicative	multiplicatieve menging
M 173	multitrace tape recorder	Mehrspurtonbandgerät n	enregistreur m à bande magnétique à pistes multiples	bandrecorder met meer geluidsporen
M 174	mush area	Interferenzgebiet n	air f d'interférence	interferentiegebied n
M 175	mushroom loudspeaker	Pilzlautsprecher m	haut-parleur m champignon	paddestoelluidspreker
M 176/7	musical ear	musikalisches Gehör n	oreille f musicale	muzikaal gehoor n
M 178	musical echo	musikalisches Echo n	écho m musical	muzikale echo
M 179	musical interval	musikalisches Intervall n	intervalle m musical	muzikaal interval n
M 180	musical pitch	musikalische Tonstufe f <in einer Tonleiter>	intervalle m musical	muzikale toonhoogte
M 181	musical scale, scale	Tonleiter f	gamme f [musicale], échelle f	toonladder
M 182	music power	Musikbelastbarkeit f <Laut-sprecher>, Musikleistung f	puissance f musicale <d'un haut-parleur>	muziekvermogen n
M 183	music wire <sl>	Saitendraht m	corde f d'instrument	snaar
M 184	mutation stop	Mutationsregister n <Orgel>	jeux m de mutation	mutatieregister n
M 185	mute	abdämpfen	assourdir, mettre en sourdine	dempen
	mute	s. a. D 6		
	muting	s. D 8		
M 186	mutual capacitance	Gegenkapazität f	capacité f mutuelle	wederzijdse capaciteit
M 187	mutual impedance	Kopplungswiderstand m	impédance f mutuelle	wederzijdse inpedantie
M 188	mutual interference	gegenseitige Störung f	interférence f réciproque	wederkerige interferentie

N

	English	German	French	Dutch
N 1	narrow acceptance-angle microphone	Schmalwinkel-Richt-mikrofon n, Mikrofon n hoher Richtschärfe	microphone m directionnel angle réduit	richtingsgevoelige micro-foon
N 2	narrow band	schmales Band n, Schmal-band n	bande f étroite	smalle band
N 3	narrow-band filter	Schmalbandfilter n	filtre m à bande étroite	selectief bandfilter n
N 4	nasal sound	Nasallaut m	son m nasal	nasaal geluid n
N 5	nasal tract	Nasenhohlraum m	fosses fpl nasales	neusholte
N 6	natural flexional frequency	Biegeeigenfrequenz f	fréquence f propre de flexion	eigenbuigingsfrequentie
N 7	natural frequency	Eigenfrequenz f	fréquence f propre	eigenfrequentie
N 8	natural interference	natürliche Störung f, Eigeninterferenz f	interférence f naturelle	natuurlijke storing
N 9	natural period	Eigenschwingungszeit f	période f propre	eigentrillingstijd
N 10	natural scatter	Eigenstreuung f	dispersion f naturelle	natuurlijke verstrooiing
	natural static	s. A 250		
N 11	natural wave	Eigenwelle f	onde f propre	natuurlijke golf
N 12	near echo	Nahecho n	écho m proche	nabije echo
N 13	near-end crosstalk	Nahnebensprechen n	diaphonie f rapprochée	nabije overspraak
	near fading	s. L 68		
N 14	near field	Nahfeld n	champ m proche	nabije veld n
N 15	necklace microphone, throat microphone, laryngophone	Kehlkopfmikrofon n	microphone m de pharynx	keelmicrofoon
N 16	needle noise	Nadelgeräusch n <Schall-platte>	bruit m d'aiguille	naaldgeruis n
N 17	negative-feedback amplifier, degenerative amplifier	gegengekoppelter Verstär-ker m	amplificateur m à contre-réaction	tegengekoppelde versterker
	negative reaction	s. C 203		
N 18	negative reaction repeater	Negistor m	négateur m	versterker met negatieve reactantie, negistor
N 19	Neper	Neper n, Np	neper m	neper
	net gain	s. Ö 45		
N 20	net resistance	Gesamtresistanz f	résistance f totale	totale weerstand
N 21	network analogue	Netzwerknachbildung f	réseau m équivalent	vervangingsschema n
N 22	network analysis	Netzwerkanalyse f	analyse f d'un réseau	netwerk-analyse
N 23	network determinant	Netzwerkdeterminante f	déterminant m d'un réseau	determinant van een netwerk
N 24	network theory	Netzwerktheorie f, Vierpol-theorie f	théorie f des réseaux (quadripôles)	netwerktheorie
N 25	neutralization of sound waves	Schallwellenkompensation f	neutralisation f d'ondes sonores	uitdoven n van geluidgolven

N 26	neutralizing capacitance	Entkopplungskapazität f	capacité f de découplage	ontkoppelingscondensator
N 27	nickelodeon <US>	elektrisches Klavier n <Münzeinwurf>	piano m électrique <à sous>	muziekautomaat
N 28	night range	Nachtreichweite f	portée f de nuit	reikwijdte bij nacht
N 29	nodal point	Knotenpunkt m	point m nodal	knoop
	node	s. O 31		
N 30	noise	Geräusch n	bruit m	geruis n
N 31	noise	Lärm m	bruit m	lawaai n, gedruis n
N 32	noise	Rauschen n	bruit m	ruis
N 33	noise abatement, noise control, acoustic control	Lärmbekämpfung f	lutte f contre le bruit, contrôle m du bruit	lawaaibestrijding
N 34	noise amplitude	Rauschamplitude f	amplitude f de bruit	ruisamplitude
N 35	noise analyzer	Geräuschanalysator m	analyseur m de bruit	geruisanalysator
N 36	noise audiogram	Geräuschaudiogramm n	audiogramme m de bruit	geruisaudiogram n
N 37	noise-cancelling microphone	störschallunterdrückendes Mikrofon n	microphone m protégé des bruits ambiants	microfoon met geruiscompensatie, storingsarme microfoon
	noise-cancelling microphone	s. a. A 218		
N 38	noise characteristic	Rauschkurve f	courbe f de bruit	ruiskarakteristiek
N 39	noise component	Rauschanteil m	composante f de bruit	ruiscomponent
N 40	noise conductance	Rauschleitwert m	conductance f de bruit	conductantie voor ruis
	noise control	s. N 33		
N 41	noise crash	Störspannungsspitze f	tension f de pointe de bruit	ruispiek
N 42	noise criteria	Lärmkriterien npl	critères mpl de bruit	geruiseisen pl
N 43	noise current	Rauschstrom m	courant m de bruit	ruisstroom
N 44	noise-exposed person	lärmexponierte Person f	personne f exposée au bruit	aan lawaai blootgestelde persoon
N 45	noise exposure	Lärmexponierung f	exposition f au bruit	lawaaibelasting, lawaaidosis
N 46/7	noise factor (figure)	Rauschfaktor m, Rauschzahl f	facteur m de bruit (qualité d'un amplificateur)	ruisfactor
N 48	noise filter, noise killer <sl>	Störschutzfilter n	filtre m de bruit	ruisfilter n
N 49	noise fluctuation	Störgeräuschatmen n	fluctuation f du bruit	ruisfluctuatie
N 50	noise-free, noiseless	rauschfrei	exempt de bruit	ruisarm
N 51	noise guards	Lärmabwehrmittel npl	protections fpl contre le bruit	lawaaibestrijdingsmiddelen npl
N 52	noise-induced hearing loss	Hörverlust m durch Lärm	perte f d'acuité auditive par bruit intense	lawaaidoofheid
N 53	noise-induced impairment of hearing	lärmbedingte Gehörbeeinträchtigung f	trouble m auditif causé par le bruit	gehoorverzwakking door lawaai
	noise intensity	s. I 88		
N 54	noise jamming	Rauschstörung f <beabsichtigte>	gêne f auditive	lawaaihinder
	noise killer <sl>	s. N 48		
N 55/6	noiseless	geräuschlos	sans bruit	geruisloos
	noiseless	s. a. N 50		
N 57	noiselessness	Geräuschfreiheit f	absence f de bruit	geruisloosheid
	noiselessness	s. a. A 2		
N 58	noise level	Geräuschpegel m	niveau m de bruit	geruispeil n, geruisniveau n
N 59	noise limitation	Störbegrenzung f	limitation f de bruit	geruisbeperking
N 60	noise-limited condition	rauschbedingte Wahrnehmungsgrenze f	condition f limitée par le bruit	door ruis beperkte waarneembaarheidsgrens
N 61	noise limiter	Störbegrenzer m	limiteur m de bruit	ruisbegrenzer
N 62	noise matching	Rauschanpassung f	adaptation f au bruit	aanpassing voor ruis
N 63	noise meter	Geräuschmeßgerät n	mesureur m de tension (niveau) de bruit	geruismeter
N 64	noise parameter	Rauschkennwert m	valeur f de bruit	ruisparameter
N 65	noise potential	Rauschspannung f	tension f de bruit	ruisspanning
N 66	noise power	Rauschleistung f	puissance f de bruit	ruisvermogen n
	noise quadripole	s. N 82		
N 67	noise rating curves	Geräuschbewertungskurven fpl; Lärmbewertungskurven fpl	courbes fpl d'estimation de bruit	grenscurven pl voor lawaai
N 68	noise reduction	Geräuschminderung f	réduction f de bruit	vermindering van geruis
N 69	noise source	Geräuschquelle f	source f de bruit	ruisbron
N 70	noise source	Lärmquelle f	source f de bruit	geruisbron
N 71	noise source	Schallquelle f	source f acoustique	bron van lawaai
N 72	noise spectrum	Rauschspektrum n	spectre m de bruit	ruisspectrum n
N 73	noise standard	Geräuschnormal n	étalon m de bruit	standaard voor geruis
N 74	noise suppressor	Lärmbegrenzer m	suppresseur m de bruit	geruisonderdrukker
N 75	noise suppressor	Geräuschbegrenzer m	suppresseur m de bruit	ruisonderdrukker
N 76	noise survey	Lärmspiegelbestimmung f	surveillance f du niveau de bruit	bepaling van het geruisniveau
N 77	noise voltage	Störspannung f	tension f de bruit	ruisspanning
N 78	noise weighting	Geräuschbewertung f; Lärmbewertung f	estimation f de bruit	weging van het geruis
N 79	noisy	geräuschvoll, lärmend	bruyant	rumoerig
N 80/1	noisy	verrauscht	troublé par le bruit	door ruis gestoord
N 82	noisy network, noise quadripole	Rauschvierpol m	quadripôle m de bruit	ruisvierpool
N 83	noisy network	rauschbehaftetes Netzwerk n	réseau m troublé	schakeling met ruisbronnen
N 84	non-attenuated	ungedämpft	non amorti	onverzwakt
N 85	non-directional	ungerichtet	non directionnel	alzijdig
N 86	non-directional microphone, omnidirectional microphone	ungerichtetes Mikrofon n, Kugelmikrofon n, Mikrofon n mit Kugelcharakteristik, Allrichtungsmikrofon n, Konferenzmikrofon n	microphone m omnidirectionnel (non directionnel)	rondom-microfoon, ongerichte microfoon
N 87	non-dissipative network	verlustloses Netzwerk n	réseau m à pertes nulles	verliesvrij netwerk n
	non-linear acoustics	s. M 1		
N 88	non-linear distortion	Linearitätsstörung f, nichtlineare Verzerrung f	distorsion f non linéaire	niet-lineaire vervorming
N 89	non-linearity	Nichtlinearität f	non-linéarité f	niet-lineariteit

N 90	non-pass attenuation	Sperrdämpfung f	atténuation f de blocage	volledige blokkering
N 91	non-reactive coupling	reaktionsfreie Kopplung f	couplage m non réactif	niet-reactieve koppeling
N 92	non-reflecting dia-phragm	nichtreflektierende Folie (Membran) f	membrane f non réfléchis-sante	reflectievrij membraan n
N 93	non-regenerative	rückkopplungsfrei	non réactif	zonder terugkoppeling
N 94	non-resonant	aperiodisch	non résonnant	niet resonerend
N 95	non-reverberant room	nachhallfreier Raum m	chambre f exempte de réverbération	galmvrije kamer
N 96	nonstationary (non-steady) noise	unstetiges Geräusch n	bruits mpl intermittents	ongelijkmatig geruis n
N 97	non-vocal consonant	stimmloser Konsonant m	consonne f sourde (muette)	stemloze medeklinker
N 98	normal auditory sensa-tion area	Normalhörfläche f	aire f d'audition normale	normale gehoorspan
N 99	normalized impact-sound level	Normtrittschallpegel m	niveau m étalon de bruit de pas	normaal contactgeluidpeil n
N 100	normalized impedance	normierte Impedanz f	impédance f étalon	normale impedantie
N 101	normalized level difference	Normalpegeldifferenz f	étalon m de différence de niveau	genormaliseerd peilverschil n
N 102	normal-mode inter-ference	Eigenschwingungsstörung f	interférence f d'onde propre	interferentie door normale trillingswijze
N 103	normal mode of vibra-tion	ungedämpfte Eigen-schwingungsmode f, Normalmode f	mode m de vibration normal	normale trillingswijze, vrije trillingswijze
N 104	normal threshold of feeling	normale Gefühlsschwelle f	seuil m normal de sensation	normale gevoeldrempel
N 105	normal threshold of hearing	Normalhörschwelle f	seuil m normal de perception auditive	normale gehoordrempel
N 106	normal threshold of pain	Normalschmerzschwelle f	seuil m normal de douleur	normale pijndrempel
N 107	notation	Notation f	notation f	notatie, notenschrift n
N 108	notch filter	Kerbfilter n	filtre m à encoches (canne-lures)	bandstopfilter n met steile flanken
N 109	null-balance amplifier	Nullabgleichverstärker m	amplificateur m à balance de zéro	nulpuntsversterker

O

O 1	objective noise meter	objektives Geräuschmeß-gerät n	appareil m de mesure objective du bruit	objectieve ruismeter
O 2	oblique pianoforte	schrägsaitiges Klavier n	piano m à cordes croisées	kruissnarige piano
O 3	obstacle gain	Hindernisgewinn m	gain m d'obstacle	versterking door een hindernis
O 4	occupational noise exposure	berufsbedingte Lärm-exponierung f	exposition f au bruit professionnelle	beroepsmatige lawaaidosis
O 5	octave band pressure level, octave pressure level	Schalldruckpegel m in einer Oktave	niveau m de pression acous-tique d'une octave	octaafpeil n, geluiddrukpeil in een octaaf
O 6	octave filter	Oktavbandpaß m	filtre m d'octave	octaaffilter n
O 7	octave flute	Pikkoloflöte f	flûte f piccolo	piccolo
	octave pressure level	s. O 5		
O 8	odd harmonic	ungeradzahlige Harmo-nische f	harmonique f impaire	oneven harmonische
O 9	odd-harmonic distortion	ungeradzahlige Verzerrung f	distorsion f d'ordre impair	vervorming door oneven harmonischen
O 10	off-channel selectivity	Trennschärfe f gegen Nachbarkanal	sélectivité f de canal adjacent	selectieve demping van naastliggende kanalen
O 11	offset angle	Neigungswinkel m	angle m d'inclinaison (de déplacement)	hellingshoek
O 12	off-tune, out-of-tune	verstimmt, falsch gestimmt	désaccordé	naast de afstemming
O 13	omnidirectional charac-teristic	Kugelcharakteristik f	caractéristique f omnidirec-tionnelle	ongerichte karakteristiek
	omnidirectional micro-phone	s. N 86		
O 14	one-degree of freedom system, single-degree of freedom system	System n mit einem Frei-heitsgrad	système m à un degré de liberté	systeem n met één vrijheidsgraad
O 15	onset	Einsatz m	attaque f	inzet
O 16	onset time	Anklingzeit f	temps m de transition	duur van de inzet
O 17	on-the-spot transmission	Direktübertragung f, Live-übertragung f, Live-sendung f	retransmission f directe	directe uitzending
O 18	open diapason	Prinzipal m ‹Orgel›	jeu m principal	hoofdwerk n
O 19	open-diaphragm loud-speaker	Lautsprecher m ohne Trichter	haut parleur m à membrane ouverte	luidspreker met open membraan
O 20	ophicleide	Ophikleide f ‹Blasinstru-ment›	ophicléide m	oficleïde
O 21	optical pattern	Lichtbandbreitenbild n ‹einer Schallplatte›	largeur f de bande optique ‹d'un disque›	glanspatroon n
O 22	optical slit alignment	optische Schlitzjustierung f	alignement m optique de fente	optische justering van de spleet
O 23	optical sound	Lichtton m	son m photoélectrique	optisch geluid n
	order of harmonics	s. H 15		
O 24	organ	Organ n, Orgel f	orgue m	orgel n
O 25	organ pipe	Orgelpfeife f	tuyau m d'orgue	orgelpijp
O 26	organ pipe scale	Faktur f ‹Orgel›	facture f	orgelpijpafmeting
O 27	organ register (stop), rank, stop	Register n ‹Orgel›	registre m d'orgue	orgelregister n
O 28	orthophony	Klangtreue f	orthophonie f, fidélité f	getrouwheid van weergave
O 29	oscillating characteristic	Schwingkennlinie f	caractéristique f d'oscillation	golvende karakteristiek
O 30	oscillation	Schwingung f	oscillation f	trilling
O 31	oscillation node, node	Schwingungsknoten m	nœud m d'oscillation	trillingsknoop
O 32	osophone	Knochenhörer m	écouteur m ostéophonique	botgeleidingstelefoon
O 33	ossicle	Gehörknöchelchen n	osselet m	gehoorbeentjes npl

O 34	osteophone	Knochenmikrofon n	microphone m ostéophonique	botgeleidingsmicrofoon
O 35	otologically normal subject	otologisch normaler Mensch m	sujet m otologiquement normal	otologisch normaal persoon
O 36	outgoing wave	austretende (ausfallende, abgestrahlte) Welle f	onde f efférente	uittredende golf
O 37	out-of-tune	s. O 12		
	output amplifier	Endverstärker m, Ausgangsverstärker m	amplificateur m de sortie	eindversterker
O 38	output level	Ausgangspegel m	niveau m de sortie	uitgangspeil n, uitgangsniveau n
O 39	output limiting selector	Ausgangsbegrenzungswahlschalter m	sélecteur m limiteur de sortie	schakelaar voor uitgangsbegrenzer
O 40	output noise ratio	Ausgangsrauschverhältnis n	rapport m de bruit de sortie	signaal-ruisverhouding aan de uitgang
O 41	output stage	Ausgangsstufe f, Endstufe f, Auskoppelstufe f	étage m de sortie	eindtrap
O 42	over-accentuation	Überbetonung f	suraccentuation f	te sterke beklemtoning
O 43	overall attenuation	Restdämpfung f	atténuation f résiduelle	totale verzwakking
O 44	overall frequency response	Gesamtfrequenzgang m	réponse f de fréquence totale	totale frequentiekarakteristiek
O 45	overall gain, net gain	Gesamtverstärkung f	gain m total d'amplification	totale versterking
O 46	overblow	überblasen <Blasinstrument>	octavier, quintoyer	overblazen
O 47	overdamping	überkritische Dämpfung f	amortissement m hypercritique	overkritische demping
O 48	overdriven amplifier	übersteuerter Verstärker m	amplificateur m sursaturé	overstuurde versterker
O 49	overlapping	Überlappung f	recouvrement m	overlapping
O 50	overload level	Übersteuerungspegel m	niveau m de sursaturation	peil n van overbelasting
O 51	overload sound pressure	Grenzschallpegel m <Mikrofon>	pression f sonore limite	geluiddruk bij overbelasting
O 52	overmatching	Überanpassung f	suradaptation f	overaanpassing
O 53	overmodulation	Übermodelung f	surmodulation f	overmodulatie
O 54	overriding of noise	Geräuschverdeckung f	couverture f du bruit	maskering van geruis
O 55	overshooting	Überschwingen n	oscillation f de dépassement	dóórschieten
O 56	overshoot ratio	Überschwingfaktor m	rapport m de dépassement	mate van dóórschieten
O 57	overtone	Oberton m	harmonique f	boventoon

<h1 style="text-align:center">P</h1>

P 1	packing of granules	Zusammenbacken n von Kohlekörnern <im Mikrofon>	agglomération f de la grenaille	samenklonteren n van kooldeeltjes
P 2	pad	s. A 256		
	pain threshold	Schmerzgrenze f, Schmerzschwellwert m	seuil m de la douleur	pijngrens
P 3	palatal sound	Gaumenlaut m	son m palatin	gehemelteklank
P 4	palate	Gaumen m	palais m	gehemelte n
P 5	palatine bone	Gaumenknochen m	os m palatin	gehemeltebeen n
	pancake loudspeaker	s. F 43		
P 6	panelling, panel lining	Wandauskleidung f	recouvrement m de la mur	wandbekleding
P 7	panoramic reception	Panoramaempfang m	réception f panoramique	panoramaontvangst
P 8	pan pipe	Panflöte f	flûte f de Pan	panfluit
P 9	parabolic-reflector microphone	Mikrofon n mit Parabolreflektor	microphone m à réflecteur parabolique	microfoon met parabolische reflector
P 10	parallel damping	Paralleldämpfung f	atténuation f parallèle	paralleldemping
P 11	parallel-pushpull amplifier	Brücken-Gegentakt-Verstärker m	amplificateur m push-pull en pont	parallele balansversterker
P 12	parallel resonance	Querresonanz f	résonance f parallèle	parallelresonantie
P 13	parallel-resonant circuit	Parallelresonanzkreis m	circuit m oscillant parallèle	parallelresonantiekring
P 14	paraphase coupling	Gegentaktschaltung f mit Phasenumkehr	couplage m antisymétrique	parafasekoppeling
P 15	parasitic oscillation	wilde Schwingung f, Parasitärschwingung f	oscillation f parasitaire	parasitair genereren
P 16	parasitic resonance	Parasitärresonanz f	résonance f parasitaire	parasitaire resonantie
P 17	parent frequency	Primärfrequenz f	fréquence f primaire	primaire frequentie
P 18	partial deafness	Schwerhörigkeit f, teilweise Taubheit f	surdité f partielle	hardhorendheid
P 19	partial node	Teilknoten m	nœud m partiel	gedeeltelijke knoop
P 20	partial noise exposure index	partieller Lärmexponierungsindex m	index m exponentiel partiel de bruit	partiële lawaaidosisindex
P 21	partial tone	Teilton m, Partialton m	ton m partiel	deeltoon
P 22	particle displacement	Schallausschlag m, Partikelauslenkung f	déplacement m de particule	deeltjesverplaatsing
P 23	particle velocity	Schallschnelle f	vélocité f d'une particule	deeltjessnelheid
	partition noise	s. I 94		
P 24	partition wall	Trennwand f	cloison f séparatrice	scheidingswand
P 25	pass band	Durchlaßbereich m	bande f passante	doorlaatband
P 26	pass band attenuation	Durchlaßdämpfung f	atténuation f de bande passante	verzwakking in de doorlaatband
P 27	pass band filter	Bandfilter n, Bandpaß m	filtre m passe-bande	bandfilter n
P 28	passive four-terminal network	passiver Vierpol m <Zweitor>	quadripôle m passif	passieve vierpool
P 29	passive sonar	Passiv-Sonar n	sonar m passif	passieve sonar
P 30	passive transducer	passiver Übertrager m	transducteur m passif	passieve transducent
P 31	path length	Weglänge f <der Übertragung>	parcours m libre, distance f de transmission	weglengte
P 32	path loss	Übertragungsdämpfung f	atténuation f de transmission	transmissieverlies n
P 33	path of propagation	Ausbreitungsweg m	voie f de propagation	voortplantingsweg, geluidpad
P 34	patter	prasseln	grésiller	knetteren, ratelen
P 35	peak	Spitzen bilden, zur Spitze anheben	monter en pointe, faire un pic	naar de top voeren

P 36	peak	Spitze *f*, Scheitel *m*	pic *m*, pointe *f*, crête *f*, sommet *m*	topwaarde
P 37	peak chopper	Schwellwertbegrenzer *m*	limiteur *m* de crête	topwaardebegrenzer
P 38	peak chopping	Spitzenbeschneidung *f*	coupure *f* des pointes	begrenzen *n* van de topwaarden
P 39	peaking	Spitzenbildung *f*, Resonanzanhebung *f*	surhaussement *m*	verhoging van de topwaarde
P 40	peak level	Spitzenpegel *m*	niveau *m* de crête	piekniveau *n*
P 41	peak noise meter	Scheitelwert-Geräuschspannungsmesser *m*	mesureur *m* de bruit crête à crête	ruispiekmeter
P 42	peak of attenuation	Dämpfungspol *m*	maximum *m* d'atténuation	maximale verzwakking
P 43	peak of oscillation	Schwingungsmaximum *n*	crête *f* d'oscillation	topwaarde van de trilling
P 44	peak of wave, wave crest	Wellenberg *m*	crête *f* d'onde	golfpiek
P 45	peak power output	Oberstrichleistung *f*	puissance *f* constante de pointe	maximaal uitgangsvermogen *n*
P 46	peak program meter	Aussteuerungsmesser *m*	indicateur *m* de profondeur de modulation	piekaanwijzende modulatiemeter
P 47	peak sound pressure	Spitzenschalldruck *m*	pression *f* sonore maximum	piekgeluiddruk, maximum geluiddruk
P 48	peak speech power	maximale Sprachleistung *f*	puissance *f* modulée de pointe	maximum spraakvermogen *n*
P 49	peak-to-peak value	Spitze-Spitze-Wert *m* ⌐ *m*	valeur *f* de pointe à pointe	top-topwaarde
P 50	peak value	Spitzenwert *m*, Scheitelwert *m*	valeur *f* de pointe	topwaarde, piekwaarde
P 51	peal	Läuten *n*	appel *m* de cloche	klokgeluid *n*, klokkenspel *r*
P 52	pedal keyboard	Pedalklaviatur *f*	pédalier *m* <d'un orgue>	pedaal *n*
P 53	pencil of sound beam	Schallstrahlenbündel *n*	faisceau *m* sonore	smalle geluidbundel
P 54	pencil transmitter	Walzenmikrofon *n*	microphone *m* à crayon	scherp gerichte zender
P 55	penetration	Durchdringung *f*	pénétration *f*	doordringen
P 56	penetration depth	Eindringtiefe *f*	profondeur *f* de pénétration	indringdiepte
P 57	perceived noise level, PNL	Lärmstärke *f*, Lästigkeitspegel *m*, ,,perceived noise level''	niveau *m* de bruit perçu	geruissterkte
P 58	percentage modulation	prozentuale Modulation *f*, Modelungsgrad *m*	profondeur *f* de modulation en pour cent	modulatiepercentage *n*, modulatiediepte in procenten
P 59	percent consonant articulation	prozentuale Konsonantenverständlichkeit *f*	pourcentage *m* de compréhensibilité des consonnes	procentuele verstaanbaarheid van medeklinkers
P 60	percent hearing	prozentuales Hörvermögen *n*	pourcentage *m* d'acuité auditive	procentueel gehoor *n*
P 61	percent hearing loss, percent impairment of hearing	prozentualer Hörverlust *m*	diminution *f* relative de l'acuité auditive en pourcent	gehoorverliesindex
P 62	perceptibility	Wahrnehmbarkeit *f*	perceptibilité *f*	waarneembaarheid
P 63	perceptible perception	wahrnehmbar *s.* D 60	perceptible	waarneembaar
P 64	perception of loudness	Lautstärkeempfindung *f*	perception *f* de la puissance sonore	waarneming van luidheid
P 65	perception of pitch, pitch perception	Tonhöhenempfindung *f*	impression *f* (perception *f*) de la hauteur de son	waarneming van toonhoogte
P 66	percussion instrument	Schlaginstrument *n*	instrument *m* à percussion	slagwerk *n*
P 67	periodic quantity	periodische Größe *f*	grandeur *f* d'oscillation	periodieke grootheid
P 68	periodic wave	periodische Welle *f*	onde *f* périodique	periodieke golf
P 69	period of oscillation	Schwingungsdauer *f*, Schwingungsperiode *f*	période *f* d'oscillation	periode van trilling
P 70	permanent hearing loss	bleibender Hörverlust *m*	perte *f* permanente d'acuité auditive	permanent gehoorverlies *n*
P 71	permanent-magnet dynamic loudspeaker	permanentdynamischer Lautsprecher *m*	haut-parleur *m* dynamique permanent	elektrodynamische luidspreker met permanente magneet
P 72	permanent-magnet erasing head	Permanentmagnet-Löschkopf *m*	tête *f* d'effacement à aimant permanent	wiskop met permanente magneet
P 73	perturbation method	Störungsmethode *f*	méthode *f* de perturbation	verstoringsmethode
P 74	phase-change coefficient	Phasenverschiebungskoeffizient *m*	coefficient *m* de déphasage	coëfficiënt van faseverschuiving
P 75	phase constant	Phasenkoeffizient *m*	constante *f* de phase	fasefactor
P 76	phase departure	Phasenabweichung *f*	différence *f* de phase	faseverschuiving
P 77	phase difference	Phasendifferenz *f*	différence *f* de phase	faseverschil *n*
P 78	phase distortion	Phasenstörung *f*	distorsion *f* de phase	fasevervorming
P 79	phase equalizer	Phasenentzerrer *m*	correcteur *m* de phase	fasecorrectienetwerk *n*
P 80	phase fading	Phasenschwund *m*	fading *m* de phase	fasefading
P 81	phase frequency distortion	Phasen-Frequenz-Störung *f*	distorsion *f* de phase-fréquence	frequentieafhankelijke fasevervorming
P 82	phase jump, rapid phase change	Phasensprung *m*	saut *m* de phase	fasesprong
P 83	phase lag, lagging of phase	Phasennacheilung *f*	retard *m* de phase, décalage *m* de phase en arrière	naijlen *n* van de fase
P 84	phase lead	Phasenvoreilung *f*	avance *m* de phase	voorijlen *n* van de fase
P 85	phase limit frequency	Phasengrenzfrequenz *f*	limite *f* de fréquence de phase	grensfrequentie voor de fase
P 86	phase response	Phasengang *m*	courbe *f* de fréquence de phase, réponse *f* en fréquence	faseresponsie
P 87	phase shift	Phasendrehung *f*, Phasenverschiebung *f*	glissement *m* de phase	fasedraaiing
P 88	phase-shift microphone	Phasenschiebemikrofon *n*	microphone *m* à glissement de phase	fasegevoelige microfoon
P 89	phase swing	Phasenhub *m*	amplitude *f* de phase	fasezwaai
P 90	phase velocity	Phasengeschwindigkeit *f*	pulsation *f* (vélocité *f*) de phase	fasesnelheid
P 91	phon	Phon *n*	phone *m*	foon
P 92	phones <sl>	Hörer *m*	écouteur *m*	hoofdtelefoon, telefoons *pl*
P 93	phonetics	Phonetik *f*	phonétique *f*	fonetiek
P 94	phonic	phonisch	phonique	fonisch, op het geluid betrekking hebbend

P 95	phonic cardiogram	Tonkardiogramm n	cardiogramme m acoustique	akoestisch cardiogram n
P 96	phonic drum	phonisches Rad n	tambour m phonique	fonisch rad n
P 97	phonic image	Lautbild n, akustisches Bild n	image f acoustique	akoestisch beeld n
P 98	phonometer	Phonometer n	phonomètre m	foonmeter
P 99	phonoscope	Phonoskop n, Schallwellen-aufzeichner m	phonoscope m	fonoscoop
P 100	phon-sone scale	Phon-sone-Skala f	échelle f phone-sone	schaal van foon tegen soon
	phrase intelligibility	s. I 84		
P 101	Phythagorean scale	pythagoräische Stimmung f	échelle f phytagoréenne	Pythagorische toonladder
P 102	pianino	Pianino n, Kleinklavier n	pianino m, petit piano m droit	pianino
P 103	pick-up	Tonabnehmer m	pick-up m, lecteur m de son	pick-up, opnemer, groef-taster
P 103a	pick-up cartridge	Tonabnehmeranker m	noyau m de pick-up	
P 104	pick-up characteristic <microphone>	Richtcharakteristik f	caractéristique f direction-nelle de microphone	richtingsdiagram n <van een microfoon>
P 105	pick-up coil	Aufnahmespule f	bobine f de pick-up	opneemspoel
P 106	piezoelectric loud-speaker	piezoelektrischer Laut-sprecher m	haut-parleur m piézo-électrique (à cristal)	piëzo-elektrische luid-spreker
	piezoelectric loud-speaker	s. a. C 240		
P 107	piezoelectric micro-phone	piezoelektrisches Mikrofon n	microphone m à cristal	piëzo-elektrische microfoon
	piezoelectric micro-phone	s. a. C 241		
P 108	piezoelectric transducer	piezoelektrischer Wandler m	transducteur m piézo-électrique	piëzo-elektrische trans-ducent
	pilot loudspeaker (monitor)	s. M 135		
P 109	pilot wave	Steuerwelle f	onde f pilote	geleidegolf
P 110	ping	Sonar-Impuls m	impulsion f de sonar	ping <van een sonar>
P 111	pink noise	Rosa-Rauschen n	perturbation f rose	rose ruis
P 112	pipe	Pfeife f, Dudelsack m	cornemuse f	fluit
P 113	piper	Dudelsackspieler m	cornemuseur m	doedelzakspeler
P 114	piston diaphragm	Kolbenmembran f	diaphragme m piston	trilplaat
P 115	pistonphone	Pistonphon n, Kolbenlaut-sprecher m	haut-parleur m à piston	pistonfoon
P 116	pitch	Tonwert m, subjektive Ton-höhe f	hauteur f de son	toonhoogte
P 117	pitch interval	Tonhöhenintervall n	intervalle m sonore	toonhoogte-interval n
P 118	pitch of tone	Tonhöhe f	hauteur f d'un son	toonhoogte
	pitch perception	s. P 65		
P 119	pitch unit	Tonhöheneinheit f, Teil-schritt m	unité f d'intervalle sonore	toonhoogtemaat
P 120	pitch-unit indicator	Tonhöhenschwankungs-messer m	indicateur m de fluctuation de fréquence sonore	indicator voor de toon-hoogte, toonhoogte-indicator
P 121	plane of polarization	Polarisationsebene f	plan m de polarisation	polarisatievlak n
P 122	plane of vibration	Schwingungsebene f	plan m d'oscillation	vlak n van trilling
P 123	plane-parallel resonator	Parallelflächenresonator m	résonateur m plan-parallèle	planparallele resonator
P 124	plane polarization	lineare Polarisation f	polarisation f linéaire	lineaire polarisatie
P 125	plane wave	ebene Welle f	onde f plane	vlakke golf
P 126	playback	Wiedergabe f, Playback m	reproduction f	weergave, naspelen n
P 127	playback characteristic	Wiedergabecharakteristik f	caractéristique f de repro-duction	weergavekarakteristiek
P 128	playback head	Wiedergabekopf m, Abspiel-kopf m	tête f de lecture	weergeefkop
P 129	playback loss	Abspielverlust m	pertes fpl de reproduction	weergeefverlies n
P 130	pleasing sound	Wohlklang m	son m agréable	aangenaam geluid n
P 131	plectrum	Plektrum n, Plektron n, Blättchen n <für Zupf-instrumente>	plectre m	plectrum n
P 132	plenum sound	Plenumklang m	son m total	plenumgeluid n
P 133	plucked string	gezupfte Saite f	corde f pincée	getokkelde snaar
P 134	plucking instrument	Zupfinstrument n	instrument m à plectre (corde pincée)	tokkelinstrument n
P 135	pneumatic loudspeaker	Druckkammerlautsprecher m, pneumatischer Laut-sprecher m	haut-parleur m à chambre de compression	pneumatische luidspreker
	PNL	s. P 57		
P 136	pocket fiddle, kid <sl>	Quartgeige f	pochette f [de maître à danser], violon m de quarte	kwartviool
P 137	point emitter	punktförmige Strahlungs-quelle f	source f ponctuelle	puntvormige straler
P 138	point impedance	Punktimpedanz f	impédance f ponctuelle	impedantie in een punt
P 139	point source	Punktquelle f <Strahler>	source f ponctuelle	puntbron
P 140	polarization filter	Polarisationsfilter n	filtre m de polarisation	polarisatiefilter
P 141	polarized sound wave	polarisierte Schallwelle f	onde f sonore polarisée	gepolariseerde geluidgolf
P 142	polar moment of inertia	polares Trägheitsmoment n	moment m d'inertie polaire	polair traagheidsmoment n
P 143	polydirectional micro-phone	Raummikrofon n, Kugel-mikrofon n	microphone m multidirec-tionnel (non directionnel)	rondom gevoelige micro-foon
P 144	polyphonic	vielstimmig, polyphon	polyphonique	polyfoon
P 145	porosity	Porosität f	porosité f	porositeit
P 146	porous absorber	poröser Absorber m	absorbant m poreux	poreuse absorber
P 147	positive feedback	Mitkopplung f	réaction f positive	terugkoppeling
P 148	post-emphasis	Nachanhebung f	renforcement m postérieur	terugduw
P 149	post-equalization	Nachentzerrung f	correction f postérieure	nacorrectie
P 150	power gain	Leistungsverstärkung f	gain m en puissance	vermogensversterking
P 151	power-handling capacity	Belastbarkeit f	charge f limite	belastbaarheid
P 152	power level	Leistungspegel m	niveau m de puissance	vermogenspeil n
P 153	power level gain	Leistungsverstärkungsmaß n	équivalent m de gain en puissance	vermogensversterkingspeil n
P 154	power spectral (spec-trum) density	spektrale Schalleistungs-dichte f	densité f du spectre de puissance	spectrale vermogensdicht-heid

	power spectrum	s. S 264		
P 155	preamplification	Vorverstärkung f	préamplification f	voorversterking
P 156	predistortion	Vorverzerrung f	prédistorsion f	vervormingscorrectie
P 157	pre-emphasis filter	Anhebungsfilter n	circuit m de préaccentuation	opduwfilter n
P 158	pre-emphasize	anheben	renforcer, accentuer	opduw
P 159	pre-equalization	Aufnahmeentzerrung f	compensation f de distorsion à l'enregistrement	opnamecorrectie
P 160	preliminary alignment	Vorabgleich m	alignement m préliminaire	voorlopige afregeling
P 161	prerecorded	vorbespielt	pré-enregistré	vooraf opgenomen
P 162	pressure doubling	Druckverdopplung f	doublement m de pression	verdubbeling van de druk
P 163	pressure gain factor	Druckgewinnfaktor m	facteur m de gain de pression	drukversterkingsfactor
P 164	pressure-gradient microphone	Druckgradientmikrofon n	capteur m de gradient de pression	drukgradiëntmicrofoon
P 165	pressure increase	Druckstauung f	croissance f de pression	drukopbouw, drukverhoging
P 166	pressure microphone	Druckmikrofon n	microphone m à caractéristique de pression	drukmicrofoon
P 167	pressure response (sensitivity)	Druckübertragungsfaktor m	facteur m de transmission de pression	drukgevoeligheid
P 168	pressure spectrum level	Pegel m des Schalldruckspektrums	niveau m du spectre de pression	spectrumpeil n van de geluiddruk
P 169	pressure transmission rating	Druckübertragungsmaß n	niveau m de transmission de pression	druktransmissiepeil n
P 170	principal axis	Bezugsachse f, Hauptachse f	axe m principal	hoofdas
P 171	principal wave	Hauptwelle f	onde f principale	voornaamste golf
P 172	print effect	Echoeffekt m, Kopiereffekt m	effet m d'écho, copie f entre couches	doordrukeffect n, echoeffect n
P 173	probability wave	Wahrscheinlichkeitswelle f	onde f de probabilité	waarschijnlijkheidsgolf
P 174	probe microphone	Sondenmikrofon n	microphone m sonde	sondemicrofoon
	progressive wave	s. A 168		
P 175	propagation	Ausbreitung f, Fortpflanzung f	propagation f	voortplanting
P 176	propagation anomaly	Ausbreitungsanomalie f	anomalie f de propagation	voortplantingsanomalie
P 177	propagation characteristic	Ausbreitungscharakteristik f	caractéristique f de propagation	voortplantingskarakteristiek
P 178	propagation coefficient	Ausbreitungskoeffizient m	coefficient m de propagation	voortplantingscoëfficiënt
P 179	propagation conditions	Ausbreitungsbedingungen fpl	conditions fpl de propagation	voortplantingscondities
P 180	propagation loss	Ausbreitungsdämpfungsmaß n	pertes fpl de propagation	voortplantingsverlies n
P 181	propagation of sound	Schallausbreitung f	propagation f du son	geluidvoortplanting
P 182	propagation of waves	Wellenausbreitung f	propagation f des ondes	golfuitbreiding, voortplanting van golven
P 183	propagation velocity, velocity of propagation	Ausbreitungsgeschwindigkeit f	vitesse f de propagation	voortplantingssnelheid
P 184	psaltery	Hackbrett n, Psalterium n	psaltérion m, tympanon m	psalterion n
P 185	psophometric electromotive force	psophometrische EMK, Geräusch-EMK f	force f électromotrice psophométrique	psofometrische elektromotorische kracht
P 186	psophometric voltage	Geräuschspannung f, psophometrische Spannung f	tension f psophométrique	psofometrische spanning
P 187	psophometric weight	Geräuschbewertungsfaktor m, psophometrischer Bewertungsfaktor m	facteur m d'évaluation psophométrique	psofometrische weegfactor
P 188	pulling range	Mitnahmebereich m	plage f d'entrainement	meeneemafstand
	pulsatance	s. C 67		
P 189	pulsating wave	pulsierende Welle f	onde f pulsatoire	pulserende golf
P 190	pulsation resonance	radialsymmetrische Resonanz f	résonance f pulsatoire	omtreksresonantie, ademresonantie
P 191	pulse bandwidth	Impulsbandbreite f	largeur f de bande d'impulsion	pulsbandbreedte
P 192	pulse dissipation	Impulsverlustleistung f	dissipation f d'impulsion	pulsdissipatie
P 193	pulse echo meter	Impulsechomeßgerät n	échomètre m d'impulsion	pulsechometer
P 194	pulse length, pulse width	Impulsdauer f, Impulslänge f	durée f d'impulsion	pulslengte
P 195	pulse rise time	Impulsanstiegszeit f, Impulssteilheit f	temps m de montée d'impulsion	flanksteilheid van de puls, pulsflanksteilheid
	pulse width	s. P 194		
P 196	pure continuous wave	reine kontinuierliche Welle f, gleichbleibende Sinuswelle f	onde f pure continue	zuivere continue golf
P 197/8	pure sound	Sinuston m, reiner Ton m	son m pur (sinusoïdal)	zuivere toon
P 199	pure tone audiometer, discrete frequency audiometer	Reintonaudiometer n	audiomètre m d'onde pure	audiometer met zuivere tonen
P 200	purity of tone	Tonreinheit f	pureté f de son	zuiverheid van toon
P 201	purling	Murmeln n	murmure m	gekabbel n
P 202	push-pull accordion	[diatonische] Handharmonika f	accordéon m diatonique	diatonische accordeon
	push-pull amplifier	s. B 20		
P 203	push-pull carbon microphone	Doppelkohlemikrofon n	microphone m à charbon push-pull	dubbelwerkende koolmicrofoon
P 204	push-pull recording	Gegentaktaufzeichnung f	enregistrement m push-pull	push-pull-opname
P 205	push-pull recording track	Gegentaktaufzeichnungsspur f	piste f d'enregistrement push-pull	push-pull-opnamespoor n
P 206	push-pull sound track	Gegentakttonspur f	piste f sonore push-pull	push-pull-geluidspoor n

Q

	Q factor	s. Q 2		
Q 1	quadratic distortion	quadratische Verzerrung f	distorsion f carrée	kwadratische vervorming
	quadripole	s. T 168		

Q 2	quality factor, Q factor	Güte *f*, Gütefaktor *m*	facteur *m* de qualité, facteur *m* Q	kwaliteitsfactor, Q-factor
Q 3	quantization noise	Quantisierungsrauschen *n*	bruit *m* de quantification	kwantiseringsruis
Q 4	quarter wave	Viertelwelle *f*	quart *m* d'onde	kwart golflengte
Q 5	quasi-plane wave	quasi-ebene Welle *f*	onde *f* quasi plane	quasi-vlakke golf
Q 6	quench frequency	Pendelfrequenz *f*	fréquence *f* d'oscillation	hikfrequentie
Q 7	quenching water	Blasenschleier *m*		smorend water *n* <door bellenscherm>
Q 8	quick fading, rapid fading	Schnellschwund *m*	fading *m* rapide	snelle fading
Q 9	quiescence	Ruhe *f*	quiétude *f*	rust
Q 10	quiet	geräuschlos	quiet, calme	stil
Q 11	quiet tuning	Stummabstimmung *f*	accord *m* silencieux	stille afstemming

R

R 1	radiate	ausstrahlen	rayonner	stralen, uitstralen
R 2	radiated noise	Schiffsgeräusch *n*, ausgestrahltes Geräusch *n*, ausgestrahlter Schall *m*	bruit *m* émis	uitgestraald geruis *n*
R 3	radiated power	Strahlungsleistung *f*	puissance *f* diffusée	uitgestraald vermogen *n*
R 4	radiated wave	ausgestrahlte Welle *f*	onde *f* diffusée	uitgezonden golf
R 5	radiating surface	Abstrahlfläche *f*, Strahleroberfläche *f*	surface *f* de radiation	stralend oppervlak *n*
R 6	radiation	Ausstrahlung *f*	radiation *f*	straling, uitstraling
R 7	radiation factor	Abstrahlgrad *m*	facteur *m* de radiation	stralingsfactor
R 8	radiation impedance	Strahlungsimpedanz *f*	impédance *f* de radiation	stralingsimpedantie
R 9	radiation index	Abstrahlmaß *n*	index *m* de radiation	stralingsindex
R 10	radiation lobe	Strahlungskeule *f*	lobe *m* de radiation	stralingsbundel
R 11	radiation sound pressure	Schallstrahlungsdruck *m*	pression *f* de radiation sonore	stralingsdruk <voor geluid>
R 12	radiogram, radio-gramophone	Phonoschrank *m*, Musiktruhe *f*, Radioapparat *m* mit Plattenspieler	ensemble *m* radio-phono	radiogrammofoon
R 13	radius of reverberation	Hallradius *m* <eines Kugelstrahlers>	rayon *m* de réflexion	galmstraal
R 14	random incidence	diffuser Schalleinfall *m*	son *m* incident diffus	diffuse inval
R 15	random-incidence response (sensitivity)	Übertragungsfaktor *m* im diffusen Schallfeld	facteur *m* de transmission en champ acoustique diffus	gevoeligheid bij diffuse inval
R 16	random noise	zufällig verteiltes Rauschen *n*	distribution *f* diffuse du bruit	ruis
R 17	range of attenuation	Dämpfungsbereich *m*	plage *f* d'atténuation	verzwakkingsbereik *n*
R 18	range of audibility	Hörbarkeitsbereich *m*	plage *f* d'audibilité	hoorbaarheidsafstand
	rank	*s.* O 27		
	rapid fading	*s.* Q 8		
	rapid phase change	*s.* P 82		
R 19	ratchet	Ratsche *f*	rochet *m*	palrad *n*
	rate of decay	*s.* D 29		
R 20	rating microphone	eingestuftes Mikrofon *n* <nach Impedanzwerten>	microphone *m* gradué	microfoon met variabele impedantie
R 21	rating sound level	Beurteilungspegel *m*	degré *m* du niveau de son	waardering van het geluidpeil
R 22	ratio amplifier	Quotientenverstärker *m*	amplificateur *m* de rapport	verhoudingsversterker
R 23	ratio of attenuation	Dämpfungsverhältnis *n*	rapport *m* d'atténuation	verzwakkingsverhouding
R 24	rattle	rasseln	faire un bruit de crécelle	geratel *n*
R 25	rattle	Schnarre *f* <Musik>	crécelle *f*	ratel
R 26	rattling noise	Rasseln *n*	bruit *m* de crécelle	ratelend geluid *n*
R 27	raw tape	Leerband *n*	bande *f* vide	lege band, maagdelijke band
R 28	Rayleigh disk	Rayleigh-Scheibe *f*	disque *m* de Rayleigh	schijf van Rayleigh
R 29	Rayleigh wave	Rayleigh-Welle *f*, Oberflächenwelle *f*	onde *f* superficielle (de Rayleigh)	Rayleigh-golf
R 30	reaction duration	Ansprechdauer *f*	durée *f* de réaction	reactietijd
R 31	reaction threshold	Ansprechschwelle *f*, Reaktionsschwelle *f*	seuil *m* de réaction	reactiedrempel
R 32	reactive attenuator	reaktives Dämpfungsglied *n*, reaktiver Abschwächer *m*	atténuateur *m* réactif	reactieve verzwakker
R 33	readability	Verständlichkeit *f* <von Signalen>	lisibilité *f*	begrijpelijkheid
R 34	readable	verständlich	lisible	begrijpelijk
R 35	realism of reproduction	Wiedergabenatürlichkeit *f*	réalisme *m* de reproduction	getrouwheid van weergave
R 36	rebounce	abprallen	rebondir	weerkaatsen
R 37	rebound	Abprall *m*	rebondissement *m*	weerkaatsing
R 38	received wave	empfangene Welle *f*	onde *f* reçue	ontvangen golf
R 39	receiver attenuation	Empfängerdämpfung *f*	atténuation *f* de récepteur	verzwakking in de ontvanger
R 40	receiver noise	Empfängerrauschen *n*	bruit *m* de fond de récepteur	ontvangerruis
R 41	receiver recovery	Empfängererholung *f*	remise *f* en état d'un récepteur	herstel *n* van de ontvanger
R 42	receiving channel	Empfangskanal *m*	canal *m* de réception	ontvangstkanaal *n*
R 43	receptivity	Aufnahmevermögen *n*	réceptivité *f*	ontvankelijkheid, receptiviteit
R 44	reciprocal of characteristic impedance	Wellenleitwert *m*	impédance *f* caractéristique réciproque	reciproke waarde van de karakteristieke impedantie
R 45	reciprocal transducer, reversible transducer	leistungssymmetrischer Übertrager *m*, reziproker Wandler *m*	transducteur *m* symétrique	reciproke transducent
R 46	reciprocal two-port network	umkehrbarer Vierpol *m*, leistungssymmetrisches Zweitor-Netzwerk *n*	quadripôle *m* symétrique	reciproke vierpool
R 47	reciprocity calibration	Reziprozitätskalibrierung *f*	calibration *f* de réciprocité	reciprociteitsijking
R 48	reciprocity coefficient	Reziprozitätskoeffizient *m*	coefficient *m* de réciprocité	reciprociteitscoëfficiënt
R 49	reciprocity principle	Reziprozitätssatz *m*	principe *m* de réciprocité	reciprociteitsprincipe *n*
R 50	recognition	Erkennung *f*	recognition *f*, reconnaissance *f*	herkenning

R 51	recognition differential	Wahrnehmbarkeitsschwelle f	seuil m de reconnaissance (différenciation)	differentiaaldrempel van herkenning
R 52	record	aufzeichnen	enregistrer	vastleggen, opnemen
R 53	record	Aufnahme f, Aufzeichnung f	enregistrement m	opname
R 54/5	record button	Aufnahmeknopf m ‹Tonbandgerät›	poussoir (bouton) m d'enregistrement	opneemtoets
R 56	recorded tape flux characteristic	Frequenzgang m des Bandflusses	caractéristique f de flux de bande	fluxkarakteristiek van de bandopname
R 57	recorder	Aufzeichnungsgerät n, Aufnahmegerät n	enregistreur m, appareil m de prise de son	opneemapparaat n, recorder
R 58	recording	Aufnahme f, Aufzeichnung f	enregistrement m	opname
R 59	recording amplifier	Aufnahmeverstärker m, Aufzeichnungsverstärker m	amplificateur m d'enregistrement	opneemversterker
R 60	recording characteristic	Aufzeichnungscharakteristik f	caractéristique f d'enregistrement	opnamekarakteristiek
R 61	recording density	Aufzeichnungsdichte f, Schreibdichte f	densité f d'écriture	opnamedichtheid
R 62	recording equalizer	Aufsprechentzerrer m, Aufzeichnungsentzerrer m, Aufnahmeentzerrer m	correcteur m de distorsion à l'enregistrement	correctienetwerk n voor opname
R 62a	recording frequency response	Aufsprechfrequenzgang m	courbe f de réponse à l'enregistrement	frequentiekarakteristiek bij opname
R 63	recording head	Tonkopf m, Aufzeichnungskopf m ⌐m	tête f enregistreuse	opneemkop
R 64	recording level meter	registrierender Pegelmesser	niveaumètre m enregistreur	uitsturingsmeter
R 65	recording loss	Aufzeichnungsverlust m	pertes fpl à l'enregistrement	opnameverlies n
R 66	recording room	Aufnahmeraum m, Studio n	studio m d'enregistrement	opnamestudio
R 67	recording slit loss	Schlitzverlust m	pertes fpl d'entrefer (de fente)	opnameverlies n door spleetbreedte
R 68	recording stylus	Aufzeichnungsstift m	style m enregistreur	snijnaald, snijbeitel
R 69	record-playback head	Sprechkopf-Hörkopf m, Aufzeichnungskopf-Wiedergabekopf m	tête f enregistreuse-liseuse	opneem-weergeefkop
R 70	record player	Plattenspieler m	phonographe m	platenspeler
R 71	record-play head	Aufnahme-Wiedergabe-Kopf m	tête f enregistreuse-liseuse	opneem-weergeefkop
R 72	record position	Aufnahmestellung f	position f d'enregistrement	opneemstand
R 73	recruitment	Recruitment n	complémentation f, restauration f, restitution f	regressie
R 74	rectangular distribution	Rechteckverteilung f	distribution f rectangulaire	rechthoekige verdeling
R 75	rectangular wave	Rechteckwelle f	onde f rectangulaire	vierkante golf
R 76	rectification efficiency	Richtwirkungsgrad m	degré m de directivité	rendement n van gelijkrichting
R 77	reduced sideband	teilweise unterdrücktes Seitenband n	bande f latérale atténuée	verzwakte zijbanden pl
R 78	reed	Zunge f ‹Frequenzmesser›	lame f ‹relais, fréquencemètre›, anche f ‹instrument›	tong ‹frequentiemeter›
R 79	reed department	Lingualstimmen fpl ‹Orgel›	registres mpl à anches	tongwerk n
R 80	reed frequency	Zungenfrequenz f	fréquence f d'anche	tongfrequentie
R 81	reed loudspeaker	Zungenlautsprecher m	haut-parleur m à lame vibrante	tongluidspreker
R 82	reed pipe	Lingualpfeife f	tuyau m à anche	tongpijp
R 83	reed stop	Lingualwerk n ‹Orgel›	jeu m à anches	tongwerkregister n
R 84	reference frequency	Bezugsfrequenz f	fréquence f de référence	referentiefrequentie
R 85	reference level	Bezugspegel m	niveau m de référence	referentiepeil n
R 86	reference quantity	Bezugsgröße f	grandeur f de référence	referentiewaarde
R 87	reference zero	Bezugsnullpunkt m	référence f de zéro	nulpunt n van de schaal
R 88	referent	Nullwert m einer Größe s. E 30	valeur f de zéro	referentiewaarde
R 89	reflected wave	Reflexion f	réflexion f	terugkaatsing, reflectie
R 90	reflection	reflexionsfrei	libre de réflexion	reflectievrij
R 91	reflectionless	Reflektorlautsprecher m	haut-parleur m à réflecteur	luidspreker met reflector
R 92	reflector loudspeaker	Reflexionsplatte f ‹eines Lautsprechers›	baffle m de réflexion	reflecterend randscherm n
R 93	reflex baffle	Baßreflexgehäuse n ‹für Lautsprecher›	reflex box m	basreflexkast
R 94	reflex box	Reflexempfang m	réception f réflexe	reflexontvangst
R 95	reflex reception	Brechungswelle f	onde f réfractée	gebroken golf
R 96	refracted wave	Brechung f	réfraction f	breking, refractie
R 97	refraction	Brechungsdämpfungsmaß n	amortissement m par réflexion	brekingsverlies n, refractieverlies n
R 98	refraction loss	Brechungsvermögen n	pouvoir m réfringent	straalbrekend vermogen n
R 99	refractive power	Schnarrwerk n ‹Orgel›, Regal n	régale m ‹jeu d'orgue›	regaal n, snorwerk n
R 100	regal	Rückkopplungsverstärkung f	amplification f régénérative	regeneratieve versterking
R 101	regenerative amplification	Klappenhorn n	grand bugle m	pistontrompet
R 102	regent's bugle	zusammenhängende Tonarten fpl	échelles fpl de tonalités	nauw verbonden toonladders pl
R 103	related scales	relativer Wahrnehmbarkeitszuwachs m, relative Reizstufe f	différence f relative de seuil	relatieve trede
R 104	relative difference limen	relative Wahrnehmbarkeitsstufe f für Frequenz	seuil m différentiel relatif de fréquence	relatieve verschildrempel van de toonhoogte
R 105	relative differential threshold of frequency	relativer Hörverlust m	perte f relative de l'ouïe	relatief gehoorverlies n
R 106	relative hearing loss	relativer Übertragungsfaktor m	sensibilité f relative en réponse	relatieve gevoeligheid
R 107	relative response (sensitivity)	Relaxationstheorie f	théorie f de la relaxation	relaxatie-theorie
R 108	relaxation theory	Relaxationszeit f	durée f de relaxation	relaxatie-tijd
R 109	relaxation time	Außenstudio n, Nebenstudio n	studio m extérieur (auxiliaire)	hulpstudio
	remote studio			

R 110	render audible	hörbar gestalten (erhalten)	rendre audible	hoorbaar maken
R 111	repeat	wiederholen	répéter	herhalen
R 112	repeat point	Spiegelabstimmpunkt *m*	point *m* de répétition	punt *n* van herhaling
R 113	repetition frequency	Folgefrequenz *f*	fréquence *f* de répétition	herhalingsfrequentie
R 114	reproduce	wiedergeben	reproduire	weergeven
R 115	reproducer	Geber *m*, Schallstrahler *m*	reproducteur *m*	weergever
R 116	reproducing head	Wiedergabekopf *m*	tête *f* de reproduction (lecture)	weergeefkop
R 117	reproducing stylus	Abspielnadel *f*	aiguille *f* de lecteur	afspeelnaald
R 118	reproduction	Wiedergabe *f*	reproduction *f*	weergave
R 119	reradiate	rückstrahlen	réfléchir	opnieuw uitstralen
R 120	reradiation attenuation	Rückstrahldämpfung *f*, Rücksprechdämpfung *f*	atténuation *f* de réflexion	verzwakking bij opnieuw uitstralen
R 121	rerecord	umschneiden <Tonband>	réenregistrer	overschrijven <geluidband>
R 122	rerecording	Umschneiden *n*	réenregistrement *m*	heropname, menging
R 123	residential noise	Lärm *m* in Wohnungen	bruit *m* d'habitations	lawaai *n* in woongebieden
R 124	residual hearing	Resthörvermögen *n*	sensibilité *f* acoustique restante	restantgehoor *n*, resterend gehoor *n*
R 125	residual hum	Restbrumm *m*	ronflement *m* résiduel	resterende brom
R 126	residual noise	Eigenrauschen *n*	bruit *m* résiduel	resterende ruis
R 127	residual sound level	Restschallpegel *m*	niveau *m* sonore résiduel	resterend-geluidpeil *n*
R 128	resistance	Resistanz *f*, Widerstand *m*	résistance *f*	weerstand
R 129	resistance-coupled amplifier	Resistanzverstärker *m*, Widerstandsverstärker *m*	amplificateur *m* à couplage par résistance	weerstandgekoppelde versterker, met weerstanden gekoppelde versterker
R 130	resistance noise	Widerstandsrauschen *n*	bruit *m* thermique (propre) de résistance	weerstandsruis
R 131	resistive potentiometer	Widerstandspotentiometer *n*	potentiomètre *m* de résistance	potentiometerweerstand
R 132	resonance	Resonanz *f*	résonance *f*	resonantie
R 133	resonance absorber	Resonanzabsorber *m*, Resonatoranordnung *f*	absorbeur *m* à résonance	resonantiedemper
R 134	resonance amplifier	Resonanzverstärker *m*, selektiver Verstärker *m*	amplificateur *m* à résonance	selectieve versterker
R 135	resonance circuit	Schwingkreis *m*, Resonanzkreis *m*	circuit *m* de résonance	resonantiekring, trillingskring
R 136	resonance frequency	Resonanzfrequenz *f*, Schwingfrequenz *f*	fréquence *f* de résonance	resonantiefrequentie
R 137	resonance peak	Resonanzspitze *f*	pointe *f* de résonance	resonantiepiek
R 138	resonance test	Klangprobe *f*	essai *m* de résonance	resonantieproef
R 139	resonance vibration	Resonanzschwingung *f*	oscillation *f* de résonance	resonantietrilling
R 140	resonant bubble	Resonanzblase *f*	bulle *f* de résonance	resonerende luchtbel
R 141	resonant cavity	Resonanzhohlraum *m*, Resonanzkammer *f*	cavité *f* de résonance	resonantieholte
R 142	resonant circuit drive	Schwingkreisanregung *f*	excitation *f* d'un circuit résonnant	aanstoten *n* van een trillingskring
R 143	resonant rise	Aufschaukeln *n* durch Resonanz	amorçage *m* par résonance	opslingering
R 144	resonate resound	mitschwingen *s.* E 12	résonner	resoneren
R 145	response	Übertragungsfaktor *m*	réponse *f*	responsie
R 146	response level	Übertragungsmaß *n*	niveau *m* (degré *m*) de transmission	responsiepeil *n*
R 147	response sensitivity	Ansprechempfindlichkeit *f*	seuil *m* de sensibilité	gevoeligheid
R 148	response to current	Stromübertragungsfaktor *m*	réponse *f* (transfert *m*) en courant	stroomgevoeligheid
R 149	response to power	Leistungsübertragungsfaktor *m*	réponse *f* en puissance, transfert *m* de puissance	vermogensgevoeligheid
R 150	response to voltage	Spannungsübertragungsfaktor *m*	réponse *f* en tension	spanningsgevoeligheid
R 151	responsiveness	Wirkmitgang *m*	responsivité *f*	aanspreekgevoeligheid
R 152	retune	nachstimmen	réaccorder	bijstemmen
R 153	return beam	Rückstrahl *m*	rayon *m* réfléchi	terugkomende straal, terugkomende bundel
R 154	return loss	Echodämpfung *f*, Echolaufzeitverlust *m*	amortissement *m* d'écho, perte *f* de réflexion	verzwakking van de terugkomende straal
R 155	reverberant absorption coefficient	Nachhallabsorptionskoeffizient *m*	coefficient *m* d'absorption de réverbération	nagalmabsorptiecoëfficiënt
R 156	reverberant field	Nachhallfeld *n*, Hall-Feld *n*	champ *m* de réverbération	nagalmveld *n*, galmveld *n*
R 157	reverberant sound	verhallter Schall *m*	son *m* réverbéré	nagalm
R 158	reverberant sound level	Nachhallpegel *m*	niveau *m* de réverbération sonore	nagalmpeil *n*
R 159	reverberate	nachhallen	réverbérer	galmen
R 160	reverberation	Nachhall *m*	réverbération *f*	nagalm
R 161	reverberation absorption coefficient	bei Nachhall gemessener Absorptionsgrad *m*	coefficient *m* d'absorption de réverbération	absorptiecoëfficiënt bij nagalm
R 162	reverberation frequency meter	Meßgerät *n* für den Frequenzgang des Nachhalls	fréquencemètre *m* de réverbération	nagalmfrequentiemeter
R 163	reverberation-limited condition	nachhallbedingte Wahrnehmungsgrenze *f*	perceptibilité *f* limitée par réverbération	door nagalm beperkte waarneembaarheidsgrens
R 164	reverberation room	Hallraum *m*	chambre *f* d'écho, chambre de réverbération	nagalmkamer
R 165	reverberation time reverse feedback	Nachhallzeit *f* *s.* D 44	temps *m* de retour d'écho	nagalmtijd
R 166	reverse phase	Gegenphase *f*	phase *f* inverse	omgekeerde fase, tegengestelde fase
	reversible transducer	*s.* R 45		
R 167	rewind	zurückspulen	rebobiner	terugwikkelen
R 168	rewind motor	Rückspulmotor *m*	moteur *m* de rebobinage	terugwikkelmotor
R 169	ribbon loudspeaker	Bändchenlautsprecher *m*	haut-parleur *m* à ruban	bandluidspreker
R 170	ribbon microphone, tape microphone	Bändchenmikrofon *n*	microphone *m* à ruban	bandmicrofoon
R 171	rich in tone	volltönend	de sonorité pleine	tonenrijk

R 172	riffle loudspeaker	Riffellautsprecher *m*	haut-parleur *m* à cannelures	riffelspeaker
R 173	right hand volume	Lautstärke *f* rechter Kanal <Stereo>	puissance *f* sonore du canal droit	volume *n* van het rechter kanaal
R 174	rigid in phase	phasenstarr	rigide en phase	fase-stabiel
R 175	ring head	Ringkopf *m* <Magnettontechnik>	tête *f* magnétique annulaire (toroïdale)	ringvormige kop
	ripple	s. 1. H 95; 2. U 33		
	ripple component	s. H 98		
R 176	ripple content	Welligkeitsanteil *m*	ronflement *m* contenu	gehalte *n* aan rimpel
R 177	ripple current	Brummstrom *m*	courant *m* de ronflement	rimpelstroom
R 178	ripple factor	Brummfaktor *m*	facteur *m* de ronflement	rimpelfactor
R 179	ripple frequency	Welligkeitsfrequenz *f*	fréquence *f* d'ondulation	rimpelfrequentie
	ripple frequency	s. a. H 101		
R 180	ripple ratio	Brummverhältnis *n*	rapport *m* de ronflement	rimpelverhouding
R 181	Rochelle salt transducer	Rochellesalzwandler *m*	transducteur *m* à cristal de Seignette	Seignettezouttransducent
R 182	room absorption	Absorptionsvermögen *n*, Absorption *f* eines Raumes	pouvoir *m* d'absorption d'une enceinte	demping in een kamer
R 183	room absorption <unit>	äquivalente Raumabsorptionsfläche *f* <Einheit>	surface *f* d'absorption équivalente <unité>	equivalent oppervlak *n* aan open raam
R 184	room acoustics, acoustics of a room	Raumakustik *f*	acoustique *f* de la salle	zaalakoestiek
R 185	room noise	Raumgeräusch *n*	bruit *m* de la salle	zaalgeruis *n*
R 186	root-mean-square sound pressure	Effektivschalldruck *m*	pression *f* sonore effective	effectieve geluiddruk
R 187	root-mean-square value	Effektivwert *m*	valeur *f* effective	effectieve waarde
R 188	rotary idler	Bandführungsrolle *f*	galet *m* de guidage	geleiderol
R 189	rotating loudspeaker	rotierender Lautsprecher *m*	haut-parleur *m* rotatif	draaiende luidspreker
R 190	rotational field	Wirbelfeld *n*	champ *m* tourbillonnaire	schuifspanningsveld *n*
R 191	rotational wave	Schubwelle *f*, Scherwelle *f*	onde *f* tourbillonnaire (de torsion)	schuifspanningsgolf
R 192	rubber ear pad	Gummimuschel *f* <Kopfhörer>	cape *f* d'écouteur en caoutchouc	rubberen oorkap
R 193	rumple	rumpeln <des Schallplattenantriebs>	bourdonner <par l'entraînement>	gestommel *n*, stommelen *n*
R 194	run-in groove	Einlaufrille *f* <Schallplatte>	sillon *m* de guidage d'entrée	inloopgroef
R 195	run-out groove	Auslaufrille *f* <Schallplatte>	sillon *m* de guidage d'arrêt	uitloopgroef
R 196	rustle	rascheln	produire un bruit de froissement	geritsel *n*

S

S 1	sabin	Sabin *n* <Einheit des Schallabsorptionsvermögens>	sabin <unité acoustique d'absorption>	sabine
S 2	Sabine absorption	Sabinesche Schallabsorption *f*	absorption *f* de Sabine	absorptie volgens Sabine
S 3	Sabine coefficient	Sabinescher Schallabsorptionsgrad *m*	coefficient *m* de Sabine	absorptiecoëfficiënt volgens Sabine
	saxhorn	s. F 52		
S 4	scale	Skale *f*, Maßstab *m*	échelle *f*	schaal
	scale	s. a. M 181		
S 5	scan	abtasten	tâter, balayer, explorer	aftasten
S 6	scanning loss	Abtastverlust *m*	pertes *fpl* de balayage	aftastverlies *n*
S 7	scatter beam	Streustrahl *m*	rayon *m* de dispersion	verstrooide straal
S 8	scattered radiation	Streustrahlung *f*	rayonnement *m* de dispersion	verstrooide straling
S 9	scattered reflection	Streureflexion *f*	réflexion *f* dispersée	verstrooide reflectie
S 10	scattering	Streuung *f*	dispersion *f*	verstrooiing
S 11	scattering cross section	Streuquerschnitt *m*	section *f* du rayon de dispersion	verstrooiingsdoorsnede
S 12	scattering cross section of a surface	Streuquerschnitt *m* einer Oberfläche	section *f* de dispersion d'une surface	verstrooiingsdoorsnede van een oppervlak
S 13	scattering loss	Streuverlust *m*	pertes *fpl* de dispersion	strooiingsverlies *n*
	scattering of sound	s. A 97		
S 14	schlieren method	Schlierenmethode *f*	méthode *f* des inhomogénéités de réfraction	Schlieren-methode
S 15	scoring system	Synchronisation *f* <beim Tonfilm>	système *m* de synchronisation	sonorisatie <van geluidfilm>
S 16	scratch filter	Geräuschfilter *n*	filtre *m* éliminateur de grattement	krasfilter *n*
S 17	scratching noise	Kratzgeräusch *n*	bruit *m* de grattement	gekras *n*
S 18	screeching sound	Kreischton *m*	son *m* criard	gegil *n*, gekrijs *n*
S 19	screen	Schirm *m*	écran *m*	scherm *n*
S 20	screened room	abgeschirmter Raum *m*	enceinte *f* blindée	afgeschermde kamer
S 21	screening	Abschirmung *f*	écran *m*, blindage *m*	afscherming
S 22	sea noise	Seegeräusch *n*, Seerauschen *n*	bruit *m* de la mer	zeeruis, geruis *n* van de zee
S 23	search tone	Suchton *m*	fréquence *f* acoustique de balayage	zoektoon
S 24	secondary tone	Nebenton *m*	fréquence *f* sonore secondaire	bijtoon, neventoon
S 25	secondary tuning	Sekundärabstimmung *f*	accord *m* secondaire	secundaire afstemming
S 26	second-order distortion	Klirrfaktor *m* 2. Ordnung	distorsion *f* carrée (de 2ème ordre)	tweedegraads vervorming
S 27	sector diaphragm	Sektormembran *f*	diaphragme *m* à secteurs	membraan *n* in sectoren
	see-as you-talk phone	s. F 1		
S 28	selectance	[dynamische] Selektion *f*	sélectance *f*	buurkanaaldemping
S 29	selective interference	selektive Störung *f*	interférence *f* sélective	selectieve interferentie
S 30	selectivity	Trennschärfe *f*	sélectivité *f*	selectiviteit
S 31	selectivity control	Selektivitätsregelung *f*, Schärfeabstimmung *f*	réglage *m* de sélectivité	regeling van de selectiviteit
S 32	self-absorption	Selbstabsorption *f*	absorption *f* propre	eigen absorptie

S 33	self-balancing	selbstabgleichend, selbst-symmetrierend	auto-symétrisant	zelfbalancering
S 34	self-excitation	Eigenerregung *f*	autoexcitation *f*	inwendige aanstoting
S 35	self-excited oscillation, self-induced oscillation, self-sustained oscillation	selbsterregte Schwingung *f*	oscillation *f* auto-excitée	inwendig opgewekte trilling
S 36	self-impedance self-induced oscillation	Eigenimpedanz *f* *s.* S 35	impédance *f* propre	inwendige impedantie
S 37	self-seeking tuner self-sustained oscillation	Automatiktuner *m* *s.* S 35	accord *m* automatique	automatisch afstemmende ontvanger
S 38	sending end	Geberende *n*, Sendeseite *f*, Signalquelle *f*	côté *m* émission, source *f* des signaux	zendzijde
S 39	sensation	Empfindung *f*	sensation *f*	gewaarwording
S 40	sensation area	Hörfläche *f*	aire *f* auditive	gehoorveld *n*
S 41	sensation level, SL	Hörpegel *m*, Empfindungspegel *m*	niveau *m* de perception	hoordrempel
S 42	sense of absolute pitch	absolutes Gehör *n*	oreille *f* absolue	absoluut gehoor *n*
S 43	sensing element	Fühler *m*, Sonde *f*	élément *m* sensoriel	gevoelig element *n*
S 44	sensitive	empfindlich	sensible	gevoelig
S 45	sensitiveness to direction	Richtungsempfindlichkeit *f*	sensibilité *f* directive	richtingsgevoeligheid
S 46	sensitivity	Empfindlichkeit *f*, Ansprechempfindlichkeit *f*	sensitivité *f*, sensibilité *f*	gevoeligheid
S 47	sensitivity control	Empfindlichkeitsregelung *f*	réglage *m* de sensibilité	gevoeligheidsregeling
S 48	sensitivity level	Übertragungsmaß *n*	niveau *m* de sensibilité	gevoeligheidspeil *n*
S 49	sensitometry	Lichtempfindlichkeitsmessung *f*, Sensitometrie *f*	sensitométrie *f*	sensitometrie
S 50	sensory perception, SP	sinnliche Wahrnehmung *f*	perception *f* sensorielle	zintuigelijke waarneming
S 51	separate	getrennt	séparé	gescheiden
S 52	series connection	Reihenschaltung *f*	connexion *f* en série	serie-schakeling
S 53	series impedance	Längsimpedanz *f*	impédance *f* série	serie-impedantie
S 54	series of pulses	Impulsklang *m*	impulsions *fpl* sérielles	reeks van pulsen
S 54a	series of sounds	Klang *m*	sons *mpl* sériels	reeks van klanken
S 55	series peaking	Reihenresonanzanhebung *f*	renforcement *m* de résonance série	opslingering door serie-resonantie
S 56	series resonance	Reihenresonanz *f*	résonance *f* série	serie-resonantie
S 57	set to music	vertonen	mettre en musique	op muziek zetten
S 58	shaded transducer	Schallwandler *m* mit modifizierter Richtcharakteristik	transducteur *m* acoustique à zone d'ombre	gewogen transducent
S 59	shading	Abblenden *n*	diaphragmation *f*, ombrage *m*	weging, controle van het richteffect
S 60	shading signal	Schattensignal *n*, Abblendsignal *n*	signal *m* d'ombre	weegsignaal *n*
S 61/2	shadow curve	Schattenkurve *f* <Audiometrie>	courbe *f* d'ombre	schaduwkromme
S 63	shadow zone	Schattenzone *f*	zone *f* d'ombre	schaduwgebied *n*
S 64	shaft sound pressure level difference	Schachtpegeldifferenz *f*, Kanalpegeldifferenz *f*, Schallpegeldifferenz *f* zwischen Nachbarschächten	différence *f* de pression sonore entre canaux	verschil *n* in geluidpeil tussen twee kanalen
S 65	sharpness of resonance	Resonanzschärfe *f*	acuité *f* de résonance	resonantiescherpte
S 66	sharp slope	steile Flanke *f*	flanc *m* raide (abrupt)	steile flank
S 67	sharp tone	scharfer Ton *m*	son *m* aigu	schelle klank
S 68	sharp tuning	Scharfabstimmung *f*	accord *m* pointu (exact, aigu)	scherpe afstemming
S 69	shawm	Schalmei *f*	chalumeau *m* <instrument de musique>, trompe *f*, trompette *f* de Martin	schalmei
S 70	shear wave	Schubwelle *f*, Scherwelle *f*	onde *f* de torsion	transversale golf
S 71	sheep gong	Schalmeiglocke *f*	clochette *f* <de mouton>	schapenbel
S 72	shield	abschirmen	blinder, protéger	afschermen
S 73	shielding factor	Schirmfaktor *m*, Abschirmfaktor *m*	facteur *m* écran	afschermingsfactor
S 74	shock	Stoß *m*	choc *m*	schok
S 75	shock excitation	Stoßerregung *f*, Stoßanregung *f*	excitation *f* par choc	stootbekrachtiging, stootexcitatie
S 76	shock insulator	Stoßisolator *m*, Körperschallisolator *m*	isolateur *m* de chocs	schokisolator
S 77	shock pulse	Stoßimpuls *m*	impulsion *f* de choc	puls
S 78	shock wave	Stoßwelle *f*	onde *f* de choc	schokgolf
S 79	short-circuit	Kurzschluß *m*	court-circuit *m*	kortsluiting
S 80	short-circuit impedance	elektrische Eingangsimpedanz *f* bei frei schwingendem System	impédance *f* de court-circuit	kortsluitimpedantie
S 81	short-distance scatter	Nahstreuung *f*	dispersion *f* proche	nabije verstrooiing
S 82	short-range fading	Nahschwund *m*	fading *m* proche (de proximité)	nabije fading
S 83	shot-effect noise	Schrotrauschen *n*	bruit *m* de la grenaille	hagelruis
S 84	shrill	schrill, grell	criard, grêle	schel
S 85	sideband attenuation	Seitenbanddämpfung *f*	atténuation *f* de bande latérale	zijbandverzwakking
S 86	sideband splash	Übersprechen *n* der Seitenfrequenzen	diaphonie *f* des fréquences latérales	zijbandoverspraak
S 87	side lobe	Nebenkeule *f*	lobe *m* latéral	zijlob, nevenmaximum *n*
S 88	side thrust	seitliche Auslenkkraft *f*, Seitendruck *m*	force *f* latérale de déviation	zijwaartse druk
S 89	sidetone	Nebenton *m*	son *m* voisin	bijtoon
S 90	signal strength	Nutzfeldstärke *f* <des Signals>	signal *m* utile	signaalsterkte
S 91	signal strength of reception	Empfangsfeldstärke *f*	puissance *f* de champ à la réception	sterkte van het ontvangstsignaal
S 92	signal-to-hum ratio	Brummabstand *m*	rapport *m* signal-ronflement	bromfactor
S 93	signal-to-interference ratio	Störabstand *m*	rapport *m* signal-interférence	stoorfactor

S 94	signal-to-noise ratio, SNR	Rauschabstand m	rapport m signal-bruit	ruisfactor, ruisgetal n
S 95	signal wave	Zeichenwelle f, zeichentragende Welle f	onde f porteuse de signaux	signaalgolf
S 96	silence	Stille f	silence m	stilte
S 97	silencer	Schalldämpfer m	silencieux m	geluiddemper
S 98	silent	still, schweigend, tonlos	silencieux	stil
S 99	silent bar	Leertakt m	silence m <notation musicale>	rust
S 100	silent zone	tote (stille) Zone f	zone f de silene	gebied n van stilte, rustgebied n
S 101	simple harmonic motion	einfache harmonische Schwingung f	oscillation f harmonique simple	eenvoudige harmonische beweging
S 102/3	simple sound source, simple source of sound	einfache Schallquelle f	source f sonore simple	eenvoudige geluidbron, niet-gerichte geluidbron
S 104	simple tone	einfacher Ton m	son m simple	enkelvoudige toon
S 105	simulation	Nachbildung f	simulation f	simulatie
S 106	simultaneity factor	Gleichzeitigkeitsfaktor m	facteur m de simultanéité	gelijktijdigheidsfactor
S 107	simultaneous reception	Simultanempfang m	réception f simultanée	simultane ontvangst
S 108	sine wave, sinusoidal wave	Sinuswelle f	onde f sinusoïdale	sinusvormige golf
S 109	singing limit	Pfeifgrenze f	limite f de sifflement	fluitgrens
S 110	singing oscillation	tönende Schwingung f	oscillation f d'amorçage	fluiten, rondzingen
S 111	singing voice	Singstimme f	voix f chantante	zangstem
S 112	single amplitude	einfache Amplitude f	amplitude f simple	enkelvoudige amplitude
S 113	single channel sound	Einkanalschall m	canal m acoustique simple	geluid n over één kanaal
	single-degree-of-freedom system	s. O 14		
S 114	single echo	Einfachecho n	écho m simple	enkelvoudige echo
S 115	single-note distortion	Eintonverzerrung f	distorsion f d'une seule note	vervorming van één toon
S 116	single sideband	Einseitenband n	simple bande f latérale	enkele zijband
S 117	single-sideband suppression	Unterdrückung f eines Seitenbandes	suppression f d'une bande latérale	onderdrukking van één zijband
S 118	single track	Einfachspur f, Einzelspur f	monopiste f, piste f unique	enkel spoor n, enkelvoudig geluidspoor n
S 119	single-track recorder	einspuriges Tonbandgerät n	enregistreur m à une piste	bandopnemer met enkelvoudig spoor
S 120	single wave	Einzelwelle f	onde f simple (unique)	enkelvoudige golf
	sinusoidal wave	s. S 108		
	site noise	s. S 393		
S 121	skew error	Zwischenkanal-Zeitfehler m <Tonband>	décalage m de temps entre trasses	scheeftrekfout <geluidband>
S 122	skew ray	schräger Strahl m, Schrägstrahl m	rayon m oblique	scheve straal
S 123	skip distance	Sprungentfernung f	distance f de saut	sprongwijdte
	SL	s. S 41		
S 124	slide trombone	Zugposaune f	trombone m à coulisse	schuiftrompet
S 125	sliding tone	gleitender Ton m	son m glissant	glijdende toon
S 126	slope	Steilheit f <der Kurve>	pente f	helling, steilheid
S 127	smoothing network	Glättungsschaltung f	circuit m de filtrage	afvlakschakeling
	SNR	s. S 94		
S 128	snubber	Stoßdämpfer m	amortisseur m <de choc>	schokdemper
S 129	soften	dämpfen, schwächen	amortir, amollir, adoucir	verzachten
S 130	soften down	abschwächen, abdämpfen	adoucir, amortir	verminderen, temperen
S 131	solid-borne sound	Körperschall m	son m émanent d'un corps solide	geluid n in vaste stoffen
	sonant consonant	s. V 54		
	sonar	s. S 214		
S 132	sonar background noise	Sonar-Fremdgeräusch n	bruit m de fond parasitant un sonar	achtergrondruis in de sonar
S 133	sonar dome	Schall[schutz]deckel m eines Übertragers	dôme m de sonar	sonardom
S 134	sonar dome insertion loss	Sonardom-Dämpfung f	pertes fpl dans le dôme d'un sonar	verzwakking door de sonardom
S 135	sonar dome loss pattern	Sonardom-Dämpfungscharakteristik f	caractéristique f d'amortissement d'un dôme de sonar	richtingsdiagram n van de sonardomverzwakking
S 136	sonar self-noise	Sonar-Eigenrauschen n	bruit m propre de sonar	sonarstoorgeruis n
S 137	sonar source level	Sonar-Sendepegel m	niveau m de source sonar	bronsterkte van de sonar
S 138	sone	Sone n	sone m	soon
S 139	sone filter	Sone-Filter n	filtre m de sone	soonfilter n
S 140	sone scale	Sone-Skala f	échelle f graduée en sone	soonschaal
S 141	sonic agglomeration	akustische Agglomeration f (Zusammenballung f)	agglomération f acoustique	opeenhoping van geluid
S 142	sonic applicator	Schallapplikator m, Beschallungsgerät n	appareil m de sonorisation	geluidapplicator, geluidkop
S 143	sonic barrier	Schallgrenze f	barrière f du son	geluidbarrière
S 144	sonic boom	Überschallknall m	bang m ultrasonique	supersonische knal
S 145	sonic cleaning	Ultraschallreinigung f	nettoyage m aux ultra-sons	reiniging door geluid
S 146	sonic degreasing	Ultraschallfettentfernung f	dégraissage m aux ultra-sons	ontvetting door geluid
S 147	sonic depth finder, supersonic echo-sounder	Ultraschallecholot m	sonde f ultrasonique	echolood n
S 148	sonic flaw detection	Fehlersuche f mit Ultraschall	détection f des défauts aux ultra-sons	akoestische detectie van materiaalfouten
S 149	sonics	Sonik f <technische Anwendung von Schallschwingungen>	sonique f	technologie van geluid, akoestische technologie
S 150	sonic soldering	akustisches Löten n	soudure f acoustique	solderen m met geluid, akoestisch solderen n
S 151	sonic surgery	akustische Chirurgie f	chirurgie f acoustique	chirurgie met geluid, akoestische chirurgie
S 152	sonic viscometry	akustische Viskositätsmessung f	viscosimétrie f acoustique	akoestische viscositeitsmeting
S 153	sonic wave	akustische Welle f	onde f acoustique	geluidgolf
S 154	sonobuoy	Sonoboje f	bouée f sonore	sonoboei
S 155	sonoluminescence	akustische Lumineszenz f	luminescence f acoustique	akoestische luminescentie

	English	German	French	Dutch
S 156	sonorous	tönend	sonore	sonoor
S 157	sonorous figures	Schwingungsfiguren *fpl*	figures *fpl* acoustiques (de Chladni)	trillingspatronen *npl*
S 158	sound	tönen, schallen	résonner	klinken
S 159	sound	Schall *m*	son *m*	geluid *n*
S 160	sound-absorbent	schallschluckend	absorbant le son	geluidabsorberend
S 161	sound absorber	Schallabsorber *m*, Schallschlucker *m*	absorbeur *m* de son	geluiddemper
S 162	sound-absorbing lining	schallschluckende Auskleidung *f*	revêtement *m* absorbant	geluidabsorberende bekleding
S 163	sound absorbing material	Schallschluckstoff *m*, Schallabsorptionsmaterial *n*	matériel *m* absorbant le son	geluidabsorberend materiaal *n*
S 164	sound-absorbing wall draping	schallschluckende Wandverkleidung *f*	recouvrement *m* mural absorbant	geluidabsorberende wandbekleding
S 165	sound absorption	Schallabsorption *f*	absorption *f* du son	geluidabsorptie
S 166	sound absorption coefficient, sound-power absorption coefficient, acoustical absorption coefficient	Schallabsorptionsgrad *m*, Schallschluckgrad *m*	pouvoir *m* d'absorption du son, coefficient *m* d'absorption acoustique	geluidabsorptiecoëfficiënt, akoestische absorptiecoëfficiënt
S 167	sound amplification	Tonverstärkung *f*	amplification *f* du son	geluidversterking
S 168	sound analyzer	Klanganalysator *m*	analyseur *m* de son	geluidanalysator
S 169	sound articulation	Lautverständlichkeit *f*	articulation *f* des sons	articulatie
S 170	sound attenuation	Schalldämpfung *f*	atténuation *f* du son	geluidverzwakking, akoestische verzwakking
	sound balancer	*s.* A 286		
S 171	sound bandwidth	Tonbandbreite *f*	largeur *f* de bande sonore	audiobandbreedte
	sound beam	*s.* A 126		
S 172	sound blend	Klangverschmelzung *f*	mélange *m* de sons	vermenging van klanken
	soundboard	*s.* S 199		
S 173	sound booth	Tonkabine *f*	cabine *f* acoustique	geluidcabine
S 174	sound box	Tonarmkopf *m*	tête *f* de pick-up	pick-upkop, geluidopnemer
S 175	sound carrier	Tonträger *m*	porteur *m* de son	geluiddrager
S 176	sound channel	Schallkanal *m*	canal *m* sonore	geluidkanaal *n*
S 177	sound column	Tonsäule *f*, Säulenlautsprecher *m*	colonne *f* de haut-parleurs	geluidzuil, luidsprekerkolom
S 178	sound communication	akustische Verbindung *f*, Tonverbindung *f*, Sprechverbindung *f*	communication *f* acoustique	geluidverbinding
S 179	sound conductivity	Schalleitfähigkeit *f*	conductivité *f* acoustique, conductibilité *f* sonore	akoestische geleidbaarheid
S 180	sound deadening	Schallabtötung *f*, Schalltilgung *f*	suppression *f* d'écho (du son)	geluiddempend
S 181	sound diffuser	Klangdiffusor *m*, Schalldiffusor *m*	diffuseur *m* de sons	klankverstrooier
S 182	sound disassociation	Klangzerfall *m*	dissociation *f* du son	uiteenvallen *n* van klanken
S 183	sound dissipation coefficient	Schalldissipationsgrad *m*	degré *m* de dissipation du son	coëfficiënt van akoestische dissipatie
S 184	sound distortion	Klangverzerrung *f*	distorsion *f* du son	geluidvervorming
S 185	sound energy	Schallenergie *f*	énergie *f* sonore	geluid-energie
S 186	sound energy density	Schallenergiedichte *f*	densité *f* d'énergie sonore	geluid-energiedichtheid, akoestische energiedichtheid
S 187	sound energy flux	Schalleistung *f*	flux *m* énergétique sonore	akoestische flux
S 188	sound energy flux density	Schallintensität *f*	densité *f* de flux énergétique sonore	geluidintensiteit
S 189	sound-energy flux density level	Schallintensitätspegel *m*	niveau *m* de densité de flux énergétique sonore	intensiteitspeil *n*
S 190	sound engineer	Tontechniker *m*	technicien *m* du son	geluidtechnicus
S 191	sound excitation	Schallerregung *f*	excitation *f* sonore	akoestische aanstoting
S 192	sound fading	Tonüberblendung *f*	raccordement *m* de sons	geluidovergang
S 193	sound field	Schallfeld *n*	champ *m* acoustique	geluidveld *n*
S 194	sound filter	Klangsieb *n*	filtre *m* acoustique	akoestisch filter *n*, klankfilter *n*
S 195	sound focussing	Schallbündelung *f*	focusation *f* du son	focussering van geluid
S 196	sound generator	Schallgeber *m*, Schallerzeuger *m*, Tongenerator *m*	générateur *m* acoustique	geluidopwekker
S 197	sound head	Tonkopf *m*	tête *f* magnétique	toonkop
S 198	sound-impression change	Klangbildveränderung *f*	changement *m* de l'image sonore	verandering van geluidindruk, verandering van klankbeeld
S 199	sounding board, sound-board	Resonanzboden *m*	table *f* de résonance	klankbodem
S 200	sound insulation <of a partition>	Schalldämmung *f* <einer Trennwand>	isolation *f* sonore <d'une cloison>	geluidisolatie
S 201	sound insulation factor of flexural waves	Biegewellendämmzahl *f*	facteur *m* d'isolement d'ondes sonores infléchies	geluidisolatie voor buiggolven
S 202	sound intelligibility	Lautverständlichkeit *f*	intelligibilité *f* des sons	verstaanbaarheid
S 203	sound intensity, sound-power intensity	Schallintensität *f*	intensité *f* du son	geluidintensiteit
S 204	sound irradiation	Beschallung *f*, Schalleinstrahlung *f*	irradiation *f* sonore	bestraling met geluid
S 205	sound isolation between rooms	Schallpegeldifferenz *f* zwischen Räumen	isolation *f* acoustique entre salles	geluidisolatie tussen kamers
S 206	sound level above threshold	überschwelliger Schall *m*	niveau *m* sonore dépassant le seuil	geluidpeil *n* boven drempel
S 207	sound level control	Schallpegelsteuerung *f*	réglage *m* de niveau sonore	regeling van het geluidpeil
S 208	sound level distribution	Schallpegelverteilung *f*	répartition *f* de niveau sonore	verdeling van het geluidpeil
S 209	sound level meter	Schallpegelmesser *m*	indicateur *m* de niveau acoustique, acoustomètre *m*	geluidpeilmeter
S 210	sound locating	Schallortung *f*	localisation *f* acoustique	akoestische plaatsbepaling

S 211	sound locator	Schallortungsgerät n	localisateur m acoustique	geluidbronlocalisator
S 212	sound mirror	Schallspiegel m, akustischer Spiegel m	miroir m acoustique	akoestische reflector
S 213	sound-modulated wave	tonmodulierte Welle f	onde f modulée par fréquences acoustiques	met geluid gemoduleerde golf
S 214	sound navigation and ranging equipment, sonar	Sonar n	sonar m	sonar
S 215	sound-on-sound record	Doppelbespielung f	enregistrement m double	dubbele geluidregistratie
S 216	sound overshooting	Übersteuerung f des Tones	surmodulation f	overmodulatie
S 217	sound particle displace-ment	Teilchenauslenkung f, Schallausschlag m	déplacement m de particule sonore	deeltjesverplaatsing
	sound particle velocity	s. S 251		
S 218	sound-path length	Schallweglänge f	distance f de déplacement acoustique	lengte van het geluidpad, akoestische weglengte
	sound perception	s. A 113		
S 219	sound permeability	Schalldurchlässigkeit f	perméabilité f acoustique	doorlaatbaarheid voor geluid, akoestische door-laatbaarheid
S 220	sound power	Schalleistung f	puissance f sonore	geluidvermogen n, akoes-tisch vermogen n
	sound-power absorption coefficient	s. S 166		
	sound-power intensity	s. S 203		
	sound-power reflection coefficient	s. S 241		
S 221	sound power through a surface element	Schalleistung f durch ein Oberflächenelement	puissance f sonore à travers un élément de surface	akoestisch vermogen n door een oppervlakte-element
S 222	sound pressure, acoustic pressure	Schalldruck m	pression f sonore (acous-tique)	geluiddruk
S 223	sound pressure level, SPL	Schalldruckpegel m	niveau m de pression sonore	geluiddrukpeil n
S 224	sound pressure micro-phone	Schalldruckmikrofon n	microphone m de pression acoustique	drukmicrofoon
S 225	sound pressure reflection coefficient	Schallreflexionsfaktor m	coefficient m de réflexion de pression acoustique	reflectiecoëfficiënt voor geluiddruk
S 226	sound probe	Schallsonde f	sonde f acoustique	geluidsonde, akoestische sonde
S 227	soundproof	schallgeschützt, schalldicht	insonorisé	geluiddicht
S 228	soundproofing	Schallschutz m	insonorisation f	echo-onderdrukking
S 229	sound propagating medium	Schallausbreitungsmedium n, Schallmedium n	milieu m de propagation du son	akoestisch medium n
S 230	sound pulse	Schallimpuls m	impulsion f sonore	geluidstoot, akoestische puls
S 231	sound radiation	Schallstrahlung f	radiation f sonore	geluiduitstraling, akoestische straling
S 232	sound radiation imped-ance	Schallstrahlungsimpedanz f	impédance f acoustique de radiation	akoestische stralingsimpe-dantie
S 233	sound radiation resist-ance	Schallstrahlungsresistanz f	résistance f à la radiation sonore	akoestische stralingsweer-stand
S 234	sound range	Schallumfang m	étendue f sonore	geluidmeetbaan
S 235	sound ranging micro-phone	Schallmeßmikrofon n	microphone m de mesure	meetmicrofoon
	sound reception	s. A 127		
S 236	sound recording	Schallaufzeichnung f	enregistrement m acous-tique	geluidregistratie
S 237	sound recording system	Tonaufnahmesystem n	système m d'enregistrement acoustique	systeem n voor geluid-registratie, geluidopneem-systeem n
S 238	sound reduction factor	Schallreduktionsfaktor m, Schalldämmverhältnis n	facteur m de réduction du son	akoestische reductie-coëfficiënt
S 239/40	sound reduction index <of a partition>	Schalldämmaß n <einer Trennfläche>	indice m de réduction du son <par une paroi>	geluidisolatie-index
S 241	sound reflection coeffi-cient, sound-power reflection coefficient	Schallreflexionsgrad m	coefficient m de réflexion sonore	akoestische reflectie-coëfficiënt
S 242	sound reproducing system	Tonwiedergabesystem n	système m de reproduction acoustique	geluidweergeefsysteem n
S 243	sound shadow	Schallschatten m	ombre f sonore	geluidschaduw, akoestische schaduw
S 244	sound source, acoustic source	Schallquelle f	source f sonore	geluidbron
S 245	sound spectrograph	Schallspektrograf m	spectrographe m acoustique	geluidspectrograaf, akoes-tische spectrograaf
S 246	sound spectrum	Schallspektrum n	spectre m sonore	geluidspectrum n
	sound supervisor	s. A 286		
S 247	sound track	Tonspur f	piste f sonore	geluidspoor n
S 248	sound transmission	Schallübertragung f	transmission f du son	geluidoverdracht, akoesti-sche transmissie
S 249	sound transmission coefficient	Schallübertragungs-koeffizient m	coefficient m de transmission du son	transmissiecoëfficiënt van het geluid, akoestische transmissiecoëfficiënt
S 250	sound tuning	Abstimmen n der Tonhöhe	accordement m des sons	stemming, stemmen n
S 251	sound velocity, acoustic velocity, sound particle velocity	Schallschnelle f	vélocité f acoustique (du son)	geluidsnelheid, deeltjes-snelheid
S 252	sound vibration, acoustic oscillation	Schallschwingung f	vibration f sonore, oscilla-tion f acoustique	geluidtrilling, akoestische trilling
S 253	sound volume	Klangfülle f	volume m sonore (de son)	geluidvolume n, akoestisch volume n
S 254	sound volume range	Dynamikbereich m des Schalls	grandeur f du volume de son	geluidvolumespan
S 255	source impedance	Quellwiderstand m, Innen-widerstand m	impédance f de source, résistance f interne	bronimpedantie
S 256	sousaphone	Sousaphon n <Tonart>	sousaphone m	Sousafoon

	SP	*s.* S 50		
S 257	space perception	räumliche Wahrnehmung *f*	perception *f* de l'espace	ruimtelijke waarneming
S 258/9	spatial effect	Raumeffekt *m*	effet *m* stéréophonique	ruimtelijk effect *n*
	speaker <US>	*s.* L 99		
S 260	speaking range	Sprechreichweite *f*	portée *f* de voix	spraakreikwijdte
S 261	speaking tube	Sprachrohr *n*, Megafon *n*	mégaphone *m*, porte-voix *m*	spreekbuis
S 262	specific acoustic imped- ance, unit-area acoustic impedance	spezifische Schallimpedanz *f*	impédance *f* acoustique spécifique	specifieke akoestische impedantie
S 263	specific acoustic mobil- ity, unit-area acoustic mobility	spezifischer Mitgang *m*	mobilité *f* acoustique spécifique	specifieke akoestische beweeglijkheid
	specific acoustic react- ance	*s.* U 43		
	specific acoustic resist- ance	*s.* U 44		
S 264	spectrum density, power spectrum	spektrale Dichte *f* <einer Feldgröße>, Leistungs- spektrum *n*	densité *f* spectrale	spectrale dichtheid
S 265	spectrum [density] level	Pegel *m* der spektralen Dichte	densité *f* spectrale de niveau sonore	spectraal peil *n*
S 266	spectrum of impact sounds	Trittschallspektrum *n*	spectre *m* d'impact sonore	contactgeluidspectrum *n*
S 267	spectrum pressure level	Schalldruckpegel *m*, bezogen auf ein Spektrum	niveau *m* spectral de pression sonore	spectraal geluiddrukpeil *n*
S 268	speech analyzer	Sprachanalysator *m*	analyseur *m* de parole	spraak-analysator
S 269	speech audiometer	Sprachaudiometer *n*	audiomètre *m* vocal	spraak-audiometer
S 270	speech band	Sprechfrequenzband *n*	bande *f* de fréquences vocales	frequentieband voor spraak
S 271	speech clipping	Sprachbeschneidung *f*, Sprachhöhenbeschnei- dung *f*	coupure *f* de fréquences vocales	amplitudebegrenzing voor spraak
S 272	speech defect	Sprachfehler *m*	défaut *m* d'élocution	spraakgebrek *n*
S 273	speech frequency	Sprachfrequenz *f*	fréquence *f* vocale	spraakfrequentie
	speech-input amplifier	*s.* M 79		
S 274	speech interference level	Sprachverständlichkeits- störpegel *m* <Mittelwert der Oktavschalldruck- pegel im mittleren Frequenzbereich>	niveau *m* d'interférence de fréquence vocale	storingspeil *n* voor spraak
S 275	speech level	Sprechpegel *m*	niveau *m* de parole	spraakpeil *n*
S 276	speech-modulated wave	sprachmodulierte Welle *f*	onde *f* modulée en fréquen- ces acoustiques	met spraak gemoduleerde golf
S 277	speech power	Sprechleistung *f*	puissance *f* vocale	spraakvermogen *n*
S 278	speech reception thresh- old, SRT	Sprachwahrnehmbarkeits- grenze *f*	limite *f* de perception de la voix	spraakwaarneembaarheids- grens
S 279	speech scrambler	Sprachverwürfler *m*		onverstaanbaarmaker van spraak, spraakverminker
S 280	speech sound	Sprechlaut *m*	son *m* vocal	spraakgeluid *n*
S 281	speech spectrogram	Sprachspektrogramm *n*	spectrogramme *m* de la parole	spraakspectrogram *n*
S 282	speech wave	Sprachwelle *f*	onde *f* vocale	spraakgolf
S 283	spherical source	Kugelstrahler *m*, kugel- förmige Schallquelle *f* (Quelle *f*)	source *f* non directionnelle	bolbron
S 284	spherical wave	Kugelwelle *f*	onde *f* sphérique	bolvormige golf
S 285	spiking	Bildung *f* von Über- schwingspitzen, Spitzen- anhebung *f*	oscillations *fpl* de dépasse- ment	optreden *n* van horentjes, uittrillen *n* na snelle veranderingen
S 286	splice	kleben <Tonband>	coller	lassen
S 287	split frequency	Spaltfrequenz *f*	fréquence *f* de fente	spleetfrequentie
S 288	split hydrophone	aufgeteiltes Hydrofon *n*, Mehrkanalhydrofon *n*	hydrophone *m* multicanaux (à fentes)	gedeelde hydrofoon
S 289	split projector	aufgeteilter Schallgeber *m*, Mehrkanalstrahler *n*	diffuseur *m* multicanaux	gedeelde projector
S 290	split transducer	Mehrkanalübertrager *m*	transducteur *m* multicanaux	gedeelde transducent
S 291	spot frequency	Rastfrequenz *f*	fréquence *f* de spot	spotfrequentie, vaste frequentie
S 292	spot noise factor	Spektralrauschfaktor *m*	facteur *m* spectral (ponctuel) de bruit	ruisfactor in de smalle band
	spreading loss	*s.* D 158		
S 293	spring suspension	federnde Aufhängung *f*, Federaufhängung *f*	suspension *f* élastique (à ressort)	verende ophanging
S 294	spurious printing	Echokopie *f*, Kopiereffekt *m*	copie *f* entre spires	doordrukeffect *n*, echo- effect *n*
S 295	squealing	Pfeifton *m*	sifflement *m*	gefluit *n*, geblaas *n*
	SRT	*s.* S 278		
S 296	stage gain	Stufenverstärkung *f*	gain *m* d'étage	versterking per trap
S 297	standard frequency	Normalfrequenz *f*	fréquence *f* étalon	standaardfrequentie
S 298	standard impact generator	Normhammerwerk *n*	générateur *m* étalon de martellement	standaardhamerapparaat *n*
S 299	standardized impact- sound	Normtrittschall *m*	bruit *m* de pas étalon	genormaliseerd contact- geluid *n*
S 300	standard microphone	Normalmeßmikrofon *n*	microphone *m* étalon (de mesure standardisé)	standaardmicrofoon
S 301	standard musical pitch <music>	Stimmtonfrequenz *f* <Musik>	fréquence *f* de son étalon	genormaliseerde muzikale toonhoogte
S 302	standard refraction	Normalbrechung *f*	réfraction *f* normalisée	genormaliseerde straal- breking
S 303	standard threshold of hearing	Standardhörschwelle *f*, Normhörschwelle *f*	seuil *m* d'audibilité normalisé	genormaliseerde gehoor- drempel
S 304/5	standard tone	Kammerton *m*	la *m* normalisé	standaardtoon
S 306	standard tuning frequency	Stimmtonfrequenz *f*	fréquence *f* normale d'accord	genormaliseerde afstemfre- quentie

	standing wave	*s.* S 309		
S 307	**static pressure**	statischer Druck *m*	pression *f* statique	statische druk
S 308	**statics**	atmosphärische Störungen *fpl*	troubles *mpl* atmosphériques	atmosferische ontladingen *pl*, atmosferische storingen *pl*
S 309	**stationary wave,** standing wave	stehende Welle *f*, Stehwelle *f*	onde *f* stationnaire	staande golf
S 310	**statistical absorption coefficient**	statistischer Schallabsorptionsgrad *m*	coefficient *m* statistique d'absorption	statistische absorptiecoëfficiënt
S 311	**steadiness of wave**	Wellenlängenkonstanz *f*	stabilité *f* (constance *f*) de longueur d'onde	golfstabiliteit
	steady sound	*s.* C 174		
S 312	**steady-state vibration**	stationäre Schwingung *f*	oscillation *f* stationnaire	stabiele trilling
S 313	**steep wave front**	steile Wellenstirnseite *f*	front *m* d'onde abrupt	steil golffront *n*
S 314	**stentorphone**	Druckluftlautsprecher *m*	stentorphone *m*, haut-parleur *m* amplificateur à air comprimé	stentorfoon
S 315	**stepped-tone jamming**	Dudeltonstörung *f*	trouble *m* par sons ondulants	meertonige storing
S 316	**step response**	Übergangsfunktion *f*, Sprungfunktion *f*	fonction *f* de transition	sprongkarakteristiek, stapresponsie
S 317	**stereo effect**	Stereoeffekt *m*	effet *m* stéréophonique	stereofonisch effect *n*
S 318	**stereo-mixer**	Stereomischpult *n*	pupitre *m* mélangeur stéréophonique	stereofonische mengtafel
	stereophonic hearing	*s.* B 127		
S 319	**stereophonic reproduction**	stereofone Wiedergabe *f*	reproduction *f* stéréophonique	stereofonische weergave
S 320	**stereophonics**	Stereofonie *f*	stéréophonie *f*	stereofonie
S 321	**stereophonic sensation**	räumliches Hören *n*	audition *f* stéréophonique	stereofonisch horen *n*
S 322	**stereophonic sound effect**	Raumtoneffekt *m*	effet *m* sonore stéréophonique	stereofonisch geluideffect *n*
S 323	**stereophonic sound system**	Stereofoniesystem *n*	système *m* stéréophonique	stereofonisch systeem *n*
S 324	**stereophonic sound system**	stereofone Übertragungsanlage *f*	équipement *m* stéréophonique	stereofonische geluidinstallatie
S 325	**stereo record changer**	Stereoplattenwechsler *m*	changeur *m* de disques stéréophoniques	stereofonische platenwisselaar
S 326	**stereo reverberation effect**	Stereonachhalleffekt *m*	effet *m* de réverbération stéréophonique	stereo-nagalmeffect *n*
S 327	**stereo sound**	Stereoton *m*	son *m* stéréophonique	stereofonisch geluid *n*
S 328	**stiffness reactance**	Steifigkeitsreaktanz *f*	réactance *f* de raideur	reactantie door stijfheid
S 329	**stimulus**	Reiz *m*	excitation *f*	prikkel
S 330	**stimulus threshold**	Reizschwelle *f*	seuil *m* d'excitation	gevoeligheidsdrempel
	stop	*s.* O 27		
S 331	**straight amplifier**	Geradeausverstärker *m*	amplificateur *m* direct	rechtuit-versterker
S 332	**straight-line nature**	Geradlinigkeit *f*	rectitude *f*, droiture *f*	rechtlijnigheid
S 333	**strain gauge**	Dehnungsmeßstreifen *m*	transducteur *m* ohmique de mesure à élongation	rekstrookje *n*
S 334	**stray coupling**	Streukopplung *f*	dispersion *f* de couplage	strooikoppeling
	street organ	*s.* B 53		
S 335	**strength of a sound source**	Schallfluß *m* einer Schallquelle	flux *m* d'une source sonore	sterkte van een geluidbron
S 336	**stress**	Druck *m*, Spannung *f*	pression *f*, contrainte *f*	spanning
S 337	**strident timbre**	helle Klangfarbe *f*	timbre *m* strident	schrille klank
S 338	**string band**	Streichorchester *n*	orchestre *m* à cordes	strijkorkest *n*
S 339	**stringed instrument**	Saiteninstrument *n*	instrument *m* à cordes	snaarinstrument *n*
S 340	**strings**	Saiten *fpl*	cordes *fpl*	snaren *pl*, strijkers *pl*
S 341	**strong coupling**	feste Kopplung *f*	couplage *m* serré	vaste koppeling, sterke koppeling
S 342	**structural resonance**	Raumresonanz *f*	résonance *f* interne	inwendige resonantie
S 343	**structure-borne sound <in buildings>**	Körperschall *m* <in Bauten>	bruit *m* propre interne <d'un bâtiment>	constructiegeluid *n*
S 344	**structure-borne sound filter**	Körperschallfilter *n*	filtre *m* de bruit interne	constructiegeluidfilter *n*
S 345	**stylus**	Nadel *f*, Abspielnadel *f*	aiguille *f*	naald
S 346	**stylus excursion**	Nadelauslenkung *f*	déviation *f* d'aiguille	naalduitwijking
S 347	**stylus force,** tracking force	Nadeldruck *m*	pression *f* d'aiguille	naalddruk
S 348	**stylus pressure control**	Auflagedruckregler *m* <Plattenspieler>	régulateur *m* de pression pour bras d'électrophone	naalddrukregelaar
S 349	**stylus sound**	Nadelton *m*	bruit *m* d'aiguille	naaldgeluid *n*
S 350	**stylus velocity**	Nadelschnelle *f*	vélocité *f* d'aiguille	naaldsnelheid
S 351	**suabe flute**	Suavial *n*	suave *f*	suabile
S 352	**subaqueous cable loudspeaker**	Unterwasserlautsprecher *m*	haut-parleur *m* subaquatique	waterdichte luidspreker
S 353	**subaudio**	infraakustisch	infra-acoustique	infra-akoestisch
S 354	**subaudio frequency**	Infraschallfrequenz *f*	fréquence *f* subaudible	infrageluidfrequentie
S 355	**subaudio wave**	Infraschallwelle *f*	onde *f* ultra-sonique	infrageluidgolf
S 356	**subharmonic**	Subharmonische *f*	subharmonique *f* <son>	subharmonische
S 357	**subharmonic response**	subharmonische Antwort *f*	réponse *f* subharmonique	subharmonische responsie
S 358	**subharmonics**	subharmonische Schwingungen *fpl*	oscillations *fpl* subharmoniques	subharmonische trillingen *pl*
S 359	**subject**	Testperson *f*, Versuchsperson *f*	sujet *m*	onderwerp *n*, proefpersoon
S 360	**subjective acoustics**	subjektive Akustik *f*	acoustique *f* subjective	subjectieve akoestiek
S 361	**subjective noise meter**	subjektives Geräuschmeßgerät *n*	mesureur *m* de tension de bruit subjectif	subjectieve geruismeter
S 362	**subjective perception**	subjektive Empfindung *f*	sensation *f* subjective	subjectieve waarneming
S 363	**subjective sound meter**	subjektiver Lautstärkepegelmesser *m*	mesureur *m* de niveau sonore subjectif	subjectieve geluidpeilmeter
S 364	**subjective tone**	subjektiver Ton *m*	son *m* subjectif	subjectieve toon
S 365	**submarine sound signal**	Unterwasserschallzeichen *n*	signal *m* sonore sous-marin	onderwatergeluidsignaal *n*
S 366	**submodulator**	Modulatorvorstufe *f*	étage *m* pré-modulateur	modulatievoortrap
S 367	**submultiple resonance**	unterharmonische Resonanz *f*	résonance *f* sous-harmonique	subharmonische resonantie
S 368	**submultiple resonance frequency**	Unterresonanzfrequenz *f*	sous-multiples *mpl* de la fréquence de résonance	subharmonische resonantiefrequentie

S 369	subrepeater	Hilfsverstärker *m*	répéteur *m* auxiliaire	tussenversterker
S 370	subsidiary wave	Nebenwelle *f*	onde *f* secondaire	nevengolf
S 371	substitute test speaker	Prüflautsprecher *m*	haut-parleur *m* de test par substitution	vervangende proefluidspreker
S 372	sum frequency	Summenfrequenz *f*	fréquence *f* intégrale	somfrequentie
S 373	summation effect	Summationseffekt *m*	effet *m* de sommation	opteleffect *n*
S 374	summation loudness	Summenlautheit *f*	somme *f* de puissance sonore	sommeringsluidheid
S 375	summation tone	Summenton *m*	son *m* de sommation	somtoon
S 376	superaudible	überhörfrequent	superaudible	boven de gehoorgrens
S 377	supercardioid microphone	Mikrofon *n* mit Supernierencharakteristik	microphone *m* supercardioïde	supercardioïde microfoon
S 378	superimpose, superpose	überlagern	superposer	superponeren
S 379	superimposition	Überlagerung *f*, Superponierung *f*	superposition *f*	superpositie
	superpose	*s.* S 378		
S 380	superregeneration	Pendelrückkopplung *f*	superrégénération *f*, superréaction *f*	superregeneratie
S 381	superregenerative network	Pendelrückkopplungsschaltung *f*	schéma *m* à superréaction	superregeneratieve schakeling
	supersonic echo-sounder	*s.* S 147		
S 382	supersonic frequency	Ultraschallfrequenz *f*	fréquence *f* ultrasonique	supersonoor
S 383	supersonic sounding	Ultraschallecholotung *f*	sondage *m* ultrasonique	supersonore dieptemeting
	supersonic wave	*s.* U 17		
S 384	superthreshold sound stimulus	überschwelliger Schallreiz *m*	excitation *f* sonore dépassant le seuil	geluidprikkel sterker dan de gevoeligheidsdrempel
S 385	suppressed carrier	unterdrückter Träger *m*	porteur *m* supprimé	onderdrukte draaggolf
S 386	suppressor	Störschutzeinrichtung *f*, Entstörvorrichtung *f*	suppresseur *m*	onderdrukker, storingsonderdrukker
S 387	supraaural earphone	Kopfhörer *m* mit ohraufliegendem (supraauralem) Kissen	écouteur *m* à coussin protecteur	tegen-het-oor-telefoon
S 388	surface backscattering differential	Oberflächenrückstreumaß *n*	rapport *m* de réflexion d'une surface de sol	terugverstrooiingscoëfficiënt van het oppervlak
S 389	surface loudspeaker	Bodenlautsprecher *m*	haut-parleur *m* de sol	ingegraven luidspreker
S 390	surface noise	Bandeigengeräusch *n*, Abspielgeräusch *n*	bruit *m* propre de bande	oppervlaktegeruis *n*
S 391	surface scattering coefficient	Oberflächenstreukoeffizient *m*	coefficient *m* de diffusion superficielle	verstrooiingscoëfficiënt van het oppervlak
S 392	surface scattering strength	Oberflächenrückstreumaß *n*	degré *m* de diffusion superficielle	verstrooiingssterkte van het oppervlak
S 393	surroundings noise, site noise	Umgebungsgeräusch *n* <Mikrofon>	bruits *mpl* d'environnement	omgevingslawaai *n*
S 394	susceptance	Suszeptanz *f*	susceptance *f*	susceptantie, terugwerkingsadmittantie
	┌ference			
S 395	susceptibility to intersustained tone	Störempfindlichkeit *f* *s.* C 174	sensibilité *f* aux interférences	storingsgevoeligheid
S 396	sweep jamming	Wobbelstörung *f*	trouble *m* par vobulation	wobbelstoring
S 397	syllable articulation	Silbenbetonung *f*, Silbenverständlichkeit *f*	articulation *f* de syllabe	lettergreepverstaanbaarheid
S 398	symmetrical transducer	symmetrischer Übertrager *m*	transducteur *m* symétrique	symmetrische transducent
S 399	sympathetic chord	mitschwingende Saite *f*	corde *f* sympathique	meetrillende snaar

T

T 1	taboo frequency	Schweigefrequenz *f*, Sperrfrequenz *f*	fréquence *f* interdite (bloquée)	verboden frequentie
T 2	take-up reel	Aufwickelspule *f*	bobine *f* d'enroulement	opwikkelspoel
T 3	talk	sprechen	parler	spreken
T 4	talk-back microphone	Gegensprechmikrofon *n*	microphone *m* de retour	terugspreekmicrofoon
T 5	talker	Sprecher *m*	speaker *m* <radio>, orateur *m*	spreker
T 6	talker echo	Sprecherecho *n*	écho *m* de parole	echo van de spreker
T 7	tank for water-borne sound measurement	Wasserschall-Meßbecken *n*	bassin *m* de mesure en acoustique hydraulique	akoestisch meetbassin *n*
T 8	tape	auf Band aufzeichnen	enregistrer sur bande	opnemen *n* op de band
T 9	tape <sl>	Tonband *n*	bande *f* magnétique	geluidband
T 10	tape background noise	Bandrauschen *n*	bruit *m* de fond de ruban magnétique	achtergrondruis van de band
T 11	tape cartridge	Bandkassette *f*	cassette *f* à ruban	geluidbandcassette
T 12	tape counter	Bandzählwerk *n*	compteur *m* de bande	bandlengte-teller
T 13	taped program	Konservenprogramm *n*	programme *m* de conserve sonore	opgenomen programma *n*
T 14	tape guide	Bandführung *f*	guide *m* de bande	bandgeleiding
T 15	tape hiss	Bandrauschen *n*	bruit *m* de bande	bandgeruis *n*
T 16	tape loop	Bandschleife *f*	boucle *f* de ruban magnétique	bandlus, band zonder eind
	tape microphone	*s.* R 170		
T 17	tape pressure	Bandandruck *m* <Tonband>, Bandschalldruck *m*	pression *f* de bande magnétique, pression *f* sonore <sur bande de micro>	banddruk
T 18	tape record	Bandaufnahme *f*	enregistrement *m* sur bande	bandopname
	tape recorder	*s.* M 15		
T 19	tape shuttle unit	Magnetbandpendeleinrichtung *f*, Bandpendeleinrichtung *f*	élément *m* oscillant pour bandes magnétiques	bandpendelapparaat *n*
T 20	tape threading	Bandeinführung *f*, Bandeinfädelung *f*	enfilage *m* du ruban	bandinleg, inleggen *n* van de band
	tapping machine	*s.* I 13		
	target strength	*s.* B 10		
T 21	telephone earphone (receiver)	Fernsprechhörer *m*	écouteur (récepteur) *m* téléphonique	telefoonhoorn, microtelefoon
T 22	telephone transmitter	Fernsprechmikrofon *n*, Sprechkapsel *f*	microphone *m*, capsule *f* téléphonique	microfoon

T 23	temporary stop button	Schnellstopptaste f	bouton m d'arrêt temporaire (instantané)	pauzetoets
T 24	temporary threshold shift, TTS	vorübergehende (zeitweise) Schwellwertverschiebung f	translation f temporaire de la valeur de seuil	voorbijgaande drempelverhoging
T 25	tension of tape	Bandzug m <Tonband>	tension f de bande	bandspanning
	terminal amplifier	s. F 36		
T 26	terminal echo suppressor	Endechosperre f	suppresseur m d'écho terminal	echo-onderdrukker
T 27	terminal resistance	Klemmwiderstand m, Klemmresistanz f, Abschlußwiderstand m	résistance f terminale	weerstand tussen de aansluitklemmen, klemweerstand
T 28	terminal velocity	Endgeschwindigkeit f	vitesse f terminale	eindsnelheid
T 29	terminating impedance	Abschlußimpedanz f	impédance f terminale	afsluitimpedantie
T 30	test tone	Prüfton m, Normalton m	ton m normal	ijktoon
T 31	theorbo	Theorbo f, Generalbaßlaute f	théorbe m, téorbe m	teorbe
T 32	thermal amplification	thermische Verstärkung f	amplification f thermique	thermische versterking
	thermal microphone	s. H 89		
T 33	thermal noise, Johnson noise	thermisches Rauschen n	bruit m (souffle m) thermique	thermische ruis
T 34	thermal tuning	thermische Stimmung f	accord m thermique	thermische afstemming
T 35	thermocline	Schicht f des maximalen Temperaturgradienten	couche f à gradient thermique maximum, thermocline f	thermische laag, thermoklien
T 36	thermophone	Thermophon n	thermophone m	thermofoon
T 37	third flute	Terzflöte f	flûte f tierce	tertsfluit
T 38	third-octave band filter	Dritteloktavbandfilter n, Terzfilter n	filtre m de bande de tiers d'octave	tertsfilter n
T 39	third-order distortion	kubische Verzerrung f	distorsion f cubique	derdegraads vervorming
T 40	three-component wave	Dreikomponentenwelle f	onde f à trois composantes	golf met drie componenten
T 41	threshold audiogram	Schwellwertaudiogramm n	audiogramme m de seuil	drempelaudiogram n
T 42	threshold of acoustic perception	Schallempfindungsschwelle f	seuil m de perception acoustique	akoestische perceptiedrempel
T 43	threshold of audibility	Hörbarkeitsschwelle f, Hörsamkeitsschwelle f	seuil m d'audibilité	hoorbaarheidsdrempel
T 44	threshold of discomfort	Störschwelle f	seuil m de gêne	onbehaaglijkheidsdrempel
T 45	threshold of feeling	Fühlschwelle f	seuil m de sensation	gevoeldrempel
T 46	threshold of hearing	Hörschwelle f	seuil m d'audition	gehoordrempel
T 47	threshold of pain	Schmerzschwelle f	seuil m de douleur	pijndrempel
T 48	threshold of speech detectibility, threshold of speech perception	Sprachwahrnehmbarkeitsschwelle f	seuil m de perception de la parole	spraakdetectiedrempel
T 49	threshold of speech intelligibility	Sprachverständlichkeitsschwelle f	seuil m d'intelligibilité de la parole	spraakverstaanbaarheidsdrempel
	threshold of speech perception	s. T 48		
T 50	threshold of speech recognition	Spracherkennbarkeitsschwelle f	seuil m de reconnaissabilité de la parole	spraakherkenningsdrempel
T 51	threshold value	Schwellenwert m	valeur f de seuil	drempelwaarde
T 52	throat loudspeaker	Kehlkopflautsprecher m	haut-parleur m de la larynx	strottehoofdluidspreker, keelluidspreker
	throat microphone	s. N 15		
T 53	throat of loudspeaker	Lautsprechertrichterhals m	cône m de haut-parleur	luidsprekernek
T 54	through wave	durchgehende Welle f	onde f continue	doorgaande golf
T 55	throw-out groove	Ausschaltrille f	sillon m d'arrêt	uitschakelgroef
T 56	tick	ticken	tiquer	tikken
T 57	tie	binden <Töne>	lier <des sons>	verbinden
T 58	tilted wave front	geneigte Wellenfront f	front m d'onde oblique	scheef golffront
T 59	timbre	Klangfarbe f	timbre m	klankkleur, timbre n
T 60	timbre change	Klangfarbenänderung f	changement m de timbre	verandering van klankkleur
T 61	timbrel	Tamburin n	tambourin m	rinkelbom, tamboerijn
T 62	tine	Zinke f <Stimmgabel>	branche f	tand van een stemvork
T 63	tinny sound	blecherner Ton m	son m de casserole	blikkerig geluid n
T 64	tonal blend	Tonmischung f	mélangeur m de sons	vermenging van tonen
T 65	tonality	Tonalität f	tonalité f	tonaliteit, toonaard
T 66	tonalizer	Klangblende f	régleur m de tonalité	tonaliteitsregelaar
T 67	tonal range	Tonumfang m	étendue f de tonalité	tonenspan
T 68	tone	[musikalischer] Ton m, Ganzton m	ton m	toon
T 69	tone arm bearing	Tonarmlager n	palier m de bras de pick-up	lager n van de pick-up-arm
T 70	tone burst	Tonimpuls m, Kurzton m	impulsion f sonore	toonpuls
T 71	tone channel	Tonkanal m	canal m acoustique	geluidkanaal n
T 72	tone control	Tonsteuerung f	réglage m de tonalité	klankkleurregeling
T 73	tone control	Tonblende f	contrôle m de tonalité	klankkleurregelaar
T 74	tone correction	Klangfarbenkorrektur f	correction f de tonalité	klankkleurcorrectie
T 75	tone filter	Tonfilter n	filtre m de tonalité	toonhoogte-filter n
T 76	tone interval	Tonintervall n	intervalle m d'un ton	toonhoogte-interval
T 77	tone perception	Tonwahrnehmung f	perception f d'un son	toonhoogtewaarneming, waarneming van toonhoogte
T 78	tone rendering	Tonwiedergabe f	reproduction f sonore	klankweergave
T 79	tone switch	Tonschalter m	commutateur m de tonalité	klankkleurschakelaar
T 80	tone tuning	Tonabstimmung f	accord m de tonalité	afstemming van de toonhoogte
T 81	tonguing	Zungenschlag m	coup m de langue	tongslag
T 82	tonic accent	Tonakzent m	accent m tonique	toonhoogteaccent n
T 83	top band	oberer Grenzwellenbereich m	bande f d'ondes supérieure	hoogste frequentieband
T 84	tormentor	Schallschluckdekor m	décor m absorbant le son	geluidabsorberend decor n
T 85	torsional vibration	Torsionsschwingung f	vibration f de torsion	torsietrilling
T 86	torsional wave	Torsionswelle f, Drehwelle f	onde f de torsion	torsiegolf
T 87	total amplitude	Gesamtamplitude f	amplitude f totale	totale amplitude
T 88	total amplitude of oscillation	Schwingungsbreite f	amplitude f totale d'oscillation	piek-piekwaarde van de trilling

T 89	**total transmission**	Totaldurchgang *m*, Gesamtdurchlaß *m*	transmission *f* totale	totale transmissie
T 90	**touch**	Anschlag *m* <Tasteninstrument>	touche *f*	aanslag
T 91	**trace**	aufzeichnen	tracer, enregistrer	neerschrijven
T 92	**tracing distortion**	Spurstörung *f*, Spurverzerrung *f*	distorsion *f* de la piste	indirect geruis *n*
T 93	**track**	Spur *f*, Tonspur *f* <Schallaufzeichnung>	piste *f*, trace *f*	spoor *n*
T 94	**tracker**	Traktur *f* <Orgel>	mutation *f*	sleep, tractuur
T 95	**tracking error** **tracking force**	Spurfehler *m* *s.* S 347	erreur *f* de la piste	snijfout
T 96	**track pitch**	Spurteilung *f*	répartition *f* des pistes	spoed van de groef
T 97	**track selection**	Spurzahl *f*	sélection *f* (numéro *m*) de la piste	spoorkeuze
T 98	**track width**	Spurbreite *f*	largeur *f* de la piste	spoorbreedte, groefbreedte
T 99	**traffic noise**	Verkehrslärm *m*	bruit *m* de trafic	verkeerslawaai *n*
T 100	**train of waves**	Wellenzug *m*	train *m* d'ondes	golftrein
T 101	**transducer**	Übertrager *m*	transducteur *m*	transducent
T 102	**transducer**	Wandler *m*, Schallaufnehmer *m*	transducteur *m*, capteur *m*	triller, opnemer
T 103	**transducer equivalent noise pressure**	äquivalenter Rauschdruck *m*	pression *f* de bruit équivalent	equivalente geluiddruk <voor transducenten ruis>
T 104	**transducer gain**	Übertragungsgewinn *m*	gain *m* de transmission	signaalwinst
T 105	**transducer loss**	Übertragungsverlust *m*	pertes *fpl* de transmission	transductieverlies *n*
T 106	**transducer sensitivity**	Wandlerempfindlichkeit *f*	sensibilité *f* de capteur	gevoeligheid voor de transducent
T 107	**transfer**	übertragen	transférer, transmettre	overdragen
T 108	**transfer**	Übertragung *f*	transmission *f*	overdracht
T 109	**transfer characteristics**	Übergangskennlinien *fpl*	caractéristiques *fpl* de transfert	overdrachtkarakteristiek
T 110	**transfer mechanical impedance**	mechanische Übertragungsimpedanz *f*	impédance *f* de transfert mécanique	mechanische-koppelingsimpedantie
T 111	**transformation**	Umwandlung *f*	transformation *f*	transformatie
T 112	**transformer**	Übertrager *m*	transformateur *m*	transformator
T 113	**transformer distortion**	Übertragungsverzerrung *f*	distorsion *f* de transformateur	transformatievervorming
T 114	**transformer loss**	Übertragerverlust *m*	pertes *fpl* de transformateur	transformatieverlies *n*
T 115	**transient oscillation**	Ausgleichsschwingung *f*	oscillation *f* de transit	sprongtrilling, uittrilling
T 116	**transient response**	Einschwingverhalten *n*, Übergangsverhalten *n*	réponse *f* de transit	sprongresponsie
T 117	**transient sound**	Übertragungston *m*, Schaltgeräusch *n*	son *m* de transmission, bruit *m* de commutation	schakelgeruis *n*
T 118	**transition**	Übergang *m*	transition *f*	uitwijking
T 119	**transition frequency**	Übergangsfrequenz *f*	fréquence *f* de transition	overgangsfrequentie
T 120	**transmissibility**	Übertragungsfaktor *m*	transmissibilité *f*	doorlaatbaarheid
T 121	**transmission band**	Durchlaßbereich *m*, Durchlaßband *n*	bande *f* de transmission	doorlaatband
T 122	**transmission channel**	Übertragungskanal *m*	canal *m* de transmission	transmissiekanaal *n*
T 123	**transmission characteristic**	Durchlaßcharakteristik *f*	caractéristique *f* de transmission	doorlaatbaarheidskarakteristiek
T 124	**transmission factor**	Übertragungsfaktor *m*	facteur *m* de transmission	doorlaatbaarheidsfactor
T 125	**transmission loss**	Schalldämmaß *n* <z. B. einer Trennwand>	pertes *fpl* de transmission	transmissieverlies *n*
T 126	**transmission loss**	Übertragungsdämpfung *f*, Übertragungsverlust *m*	pertes *fpl* de transmission	transmissieverlies *n*
T 127	**transmit, X-mit**	senden	émettre, transmettre	zenden
T 128	**transmittance**	Durchlässigkeit *f*	transmittance *f*	doorlaatbaarheid
T 129	**transmitted wave**	durchgelassene Welle *f*	onde *f* transmise	uitgezonden golf, doorgelaten golf
T 130	**transmitter, X-mitter**	Sender *m*	émetteur *m*, transmetteur *m*	zender
T 131	**transmitter**	Mikrofon *n* <in einem Fernsprechkanal>	transmetteur *m* <en téléphonie: microphone>	microfoon
T 132	**transmitting, X-mitting**	Senden *n*	émission *f*, transmission *f*	uitzending
T 133	**transmitting voltage response**	spannungsbezogener Schallübertragungskoeffizient *m*	réponse *f* de tension en pression sonore	frequentiekarakteristiek bij constante spanning
T 134	**transversal modulation**	Transversalmodulation *f*	modulation *f* transversale	transversale modulatie
T 135	**transversal vibration** **transverse flute**	Transversalschwingung *f* *s.* C 230	oscillation *f* transversale	transversale trilling
T 136	**transverse wave**	Transversalwelle *f*	onde *f* transversale	transversale golf
T 137	**travel**	Weg *m*, Auslenkung *f*	déplacement *m*, déviation *f*	afgelegde weg
T 138	**travelling wave**	Wanderwelle *f*, fortschreitende Welle *f*	onde *f* progressive (de translation)	lopende golf
T 139	**treble**	hohe Töne *mpl* (Tonfrequenzen *fpl*)	sons *mpl* hauts	sopraan . . ., hoge tonen *pl*
T 140	**treble control**	Höhenregler *m*	réglage *m* d'aiguës	hoge-tonenregeling, regeling van de hoge tonen
T 141	**treble correction**	Höhenentzerrung *f*	correction *f* d'aiguës	hoge-tonencorrectie, correctie voor de hoge tonen
T 142	**treble lift and cut**	Höhenanhebung und -beschneidung *f*	renforcement *m* des aiguës et des basses	versterking en verzwakking van hoge tonen
T 143	**treble loudspeaker**	Hochtonlautsprecher *m*	haut-parleur *m* d'aiguës	hoge-tonenluidspreker, luidspreker voor de hoge tonen
T 144	**trick button**	Tricktaste *f*	bouton *m* à effets, bouton *m* de truquage	truc
T 145	**trick network**	Kunstschaltung *f*	schéma *m* de truquage	trucage-netwerk *n*
T 146	**trim**	trimmen	trimmer	afregelen, trimmen
T 147	**trimmer**	Trimmer *m* <Kondensator>	trimmer *m*	trimmer
T 148	**trimming**	Nachstimmen *n*	ajustage *m*, alignement *m*	afregeling
T 149	**triole**	Triola *f*	triolet *m*	triool
T 150	**triplet**	Triole *f*	triolet *m*	triool
T 151	**true-to-life sound reproduction**	natürliche Tonwiedergabe *f*	reproduction *f* naturelle des sons	natuurgetrouwe geluidweergave

	TTS	*s.* T 24		
T 152	tune	Melodie *f*	mélodie *f*	melodie, wijs
T 153	tune, tune-in	stimmen, abstimmen	accorder	stemmen, afstemmen
T 154	tuned circuit	Abstimmkreis *m*	circuit *m* d'accord	afgestemde kring
T 155/6	tuned damper	abgestimmter Dämpfer *m*	atténuateur *m* accordé	resonantiedemper
	tune-in	*s.* T 153		
	tuning control	*s.* T 160		
T 157	tuning fork	Stimmgabel *f*	diapason *m*	stemvork
T 158	tuning hammer	Stimmhammer *m*	marteau *m* à accorder	stemhamer
T 159	tuning indicator	Abstimmanzeiger *m*	indicateur *m* d'accord	afstemindicator
T 160	tuning knob, tuning control	Abstimmknopf *m*	bouton *m* de réglage	afstemknop
T 161	tuning peg	Stimmwirbel *m*	cheville *f*	stemsleutel
T 162	tuning pitch	Stimmton *m*	ton *m* d'accord	stemtoon
T 163	tuning wire	Stimmdraht *m*	fil *m* d'accord	stemkruk
	turkish crescent	*s.* B 105		
T 164	turntable	Plattenteller *m*	plateau *m* de tourne-disque	draaitafel
T 165	tweeter	Hochtonlautsprecher *m*	tweeter *m*, haut-parleur *m* pour sons aigus	discantluidspreker, luidspreker voor hoge tonen
T 166	twin sound locator	Doppelrichtungshörer *m*	localisateur *m* à double son	dubbele geluidbronlocalisator
T 167	twin-track recording	Doppelspuraufzeichnung *f*	enregistrement *m* à double piste	dubbelsporige opname
	twisted loudspeaker horn	*s.* C 104		
T 168	two-port network, four-terminal network, quadripole	Vierpol *m*, Zweitor *n*	quadripôle *m*, réseau *m* biporte	vierpool
T 169	tympanic cavity	Paukenhöhle *f* ‹Ohr›	cavité *f* tympanique	trommelholte
	tympanum	*s.* E 4		

U

	ukulele	*s.* H 31		
U 1	ultra-audible frequency	Überhörfrequenz *f*	fréquences *fpl* supraaudibles	ultrageluidfrequentie
U 2	ultra-audion	Rückkopplungsaudion *n*	audion *m* à réaction	ultra-audion *n*
	ultra-audio wave	*s.* U 17		
U 3	ultra-directional microphone	Super-Richtmikrofon *n*	microphone *m* ultra-directionnel	scherp gerichte microfoon
U 4	ultrasensitive	überempfindlich	ultra-sensible	overgevoelig
U 5	ultrasonic beam	Ultraschallstrahl *m*	rayon *m* d'ultra-sons	ultrageluidbundel
U 6	ultrasonic cross grating	Ultraschallkreuzgitter *n*, Beugungsgitter *n*	réseau *m* de diffraction ultra-sonique	ultrasonoor buigingsrooster *n*
U 7	ultrasonic detector	Ultraschalldetektor *m*	détecteur *m* d'ultra-sons	ultrasonore detector
U 8	ultrasonic generator	Ultraschallgenerator *m*	générateur *m* ultra-sonique	ultrasonore generator
U 9	ultrasonic grating constant	Ultraschallbrechungskonstante *f*	facteur *m* de diffraction d'ultra-sons	ultrasonore diffractieconstante
U 10	ultrasonic light diffraction	Ultraschallichtbeugung *f*	diffraction *f* ultrasonique optique	lichtdiffractie door ultrasonore trillingen
U 11	ultrasonic receiver	Ultraschallfühler *m*	récepteur *m* d'ultra-sons	ultrasonore ontvanger
U 12	ultrasonics	Ultraschalltechnik *f*	technique *f* ultrasonique	ultrageluidtechniek
U 13	ultrasonic sounding	Ultraschallotung *f*	sondage *m* ultrasonique	ultrasonore dieptemeting
U 14	ultrasonic space grating	Ultraschallraumgitter *n*	grille *f* d'espace ultra-sonique	ruimtelijk ultrasonoor buigingsrooster *n*
U 15	ultrasonic stroboscope	Ultraschallstroboskop *n*	stroboscope *m* ultrasonique	ultrasonore stroboscoop
U 16	ultrasonic transmitter	Ultraschallübertrager *m*	transmitteur *m* d'ultra-sons	ultrasonore zender
U 17	ultrasonic wave, ultra-audio wave, supersonic wave	Ultraschallwelle *f*	onde *f* ultra-sonique	ultrageluidgolf
U 18	ultrasound	Ultraschall *m*	ultra-son *m*	ultrageluid *n*
U 19	unaudible	unhörbar	inaudible	onhoorbaar
U 20	unbalance	Unsymmetrie *f*	asymétrie *f*, dissymétrie *f*	onbalans
U 21	uncoupled mode	ungekoppelte Mode *f*	mode *m* non couplé	niet-gekoppelde trillingswijze
U 22	undamped natural frequency	ungedämpfte Eigenfrequenz *f*	fréquence *f* propre non amortie	ongedempte eigen frequentie
U 23	underwater ambient noise	Unterwasserumgebungsrauschen *n*	bruit *m* ambiant sous-aquatique	omgevingslawaai *n* onder water
U 24	underwater background noise	Unterwassergrundrauschen *n*	bruit *m* de fond sous-aquatique	achtergrondgeruis onder water
U 25	underwater echo ranging	Unterwasserentfernungsmessung *f*	télémétrage *m* sous-marin à écho sonore	onderwater-detectie
U 26	underwater self-noise	Unterwassereigenrauschen *n*	bruit *m* de fond propre sous-marin	eigen stoorgeruis *n* onder water
U 27	underwater sound	Unterwasserschall *m*	son *m* subaquatique	geluid *n* onder water, onderwatergeluid *n*
U 28	underwater sound detector	Unterwasserschallempfänger *m*	détecteur *m* de son sous-marin	onderwatergeluiddetector
U 29	underwater sound projector	Unterwasserschallstrahler *m*	émetteur *m* de sons sous-marin	onderwaterluidspreker
U 30	undesired coupling	Streukopplung *f*	couplage *m* parasite (accidentel)	ongewenste koppeling
U 31	undesired reflection	Störreflexion *f*	réflexion *f* parasitaire (accidentelle)	ongewenste reflectie
U 32	undulated	gewellt	ondulé	gegolfd
U 33	undulation, ripple	Wellenschwingung *f*, Welligkeit *f*	ondulation *f*	golfbeweging, rimpel
U 34/5	unidirectional effect	Richtwirkung *f*	effet *m* unidirectionnel	richteffect *n*
U 36	unidirectional microphone	Mikrofon *n* mit einseitiger Richtwirkung	microphone *m* unidirectionnel	eenzijdig gevoelige microfoon
U 37	uniform-spectrum random noise, white noise	weißes Rauschen *n*	bruit *m* blanc	witte ruis

U 38	unilateral transducer	einseitig wirkender Übertrager *m*, nicht umkehrbarer Übertrager *m*	transducteur *m* unilatéral	eenzijdige transducent
U 39	unison	im Gleichklang	unisson	gelijkluidend, unisono
U 40	unisonance	Gleichklang *m*	unisson *f*	gelijkluidendheid
U 41	unisonant	gleichtönend	unissonnant, unisson	gelijkluidend
U 42	unison stops	Äqualregister *n* ‹Orgel›	unisson *f* ‹registre d'orgue›	achtvoetregister *n* ‹orgel›
	unit-area acoustic impedance	*s.* S 262		
	unit-area acoustic mobility	*s.* S 263		
U 43	unit-area acoustic reactance, specific acoustic reactance	spezifische Schallreaktanz *f*	réactance *f* acoustique spécifique	specifieke akoestische reactantie
U 44	unit-area acoustic resistance, specific acoustic resistance	spezifische Schallresistanz *f*	résistance *f* acoustique spécifique	specifieke akoestische weerstand
U 45	unreadable	unverständlich ‹Funksignal›	incompréhensible	onverstaanbaar
U 46	unresponsiveness	Wirkstandwert *m*	non-responsivité *f*	aanspreekongevoeligheid
U 47	untuned	unabgestimmt	non accordé	niet afgestemd
U 48	unwanted sound ‹of audiometer›	Eigenstörungen *fpl* ‹des Audiometers›	bruit *m* de fond ‹d'un audiomètre›	stoorgeruis *n*
U 49	unweighted	unbewertet	non apprécié, non pondéré (étalonné)	niet gewogen
U 50	unweighted signal-to-noise ratio	Fremdspannungsabstand *m*	rapport *m* signal-bruit non étalonné	rauwe signaal-ruisverhouding
U 51	upper partial	Oberton *m*, oberer Teilton *m*	harmonique *f* supérieure	boventoon
U 52	upward modulation	additive Modulation *f*	modulation *f* additive	additieve modulatie
	usable field	*s.* U 53		
U 53	use field, usable field	Nutzfeld *n*	champ *m* utilisable	bruikbare veld *n*

V

V 1	valve hum	Röhrenbrumm *m*	ronflement *m* de la valve redresseuse	buisbrom, gloeispanningsbrom
V 2	valve noise	Röhrenrauschen *n*	bruit *m* de la tube, souffle *m*	buisruis
V 3	variable-area track	Zackenschrifttonspur *f*	piste *f* de la gravure à élongation variable	transversaalsysteem *n*, zaagtandsysteem *n*
V 4	variable attenuator	Dämpfungsregler *m*, veränderliches Dämpfungsglied *n*	atténuateur *m* réglable	variabele verzwakker
V 5	variable-density track	Sprossenschrifttonspur *f*, Tonspur *f* mit veränderlicher Dichte	piste *f* à densité variable	geluidspoor *n* met variabele dichtheid, intensiteitssysteem *n*
V 6	variable-inductance pick-up	Tonabnehmer *m* mit veränderlicher Induktivität	lecteur *m* à inductivité variable	pick-up met variabele inductantie
V 7	variable noiseless recording	[stellbare] Klarton-Amplitudenaufzeichnung *f*	enregistrement *m* à sonorité épurée variable	ruisvrije opname
V 8	velocity	Schnelle *f*, [translatorische] Geschwindigkeit *f*	vélocité *f*	snelheid
V 9	velocity amplitude	Schnellamplitude *f*, Geschwindigkeitsamplitude *f*	amplitude *f* de la vélocité	snelheidsamplitude
V 10	velocity field	Schnellefeld *n*	champ *m* de la vélocité	snelheidsveld *n*
V 11	velocity level	Schnellepegel *m*	niveau *m* de la vélocité	snelheidspeil *n*
V 12	velocity meter	Schnellemesser *m*	vélocimètre *m*, mesureur *m* de vélocité	snelheidsmeter
V 13	velocity microphone	Schnellemikrofon *n*, Geschwindigkeitsmikrofon *n*	capteur *m* de la gradient de vitesse	snelheidsmicrofoon, drukgradiëntmicrofoon
V 14	velocity modulation	Schnellemodulation *f*, Geschwindigkeitsmodulation *f*	modulation *f* de la vitesse	snelheidsmodulatie
	velocity of propagation	*s.* P 183		
V 15	velocity pick-up	Geschwindigkeitsaufnehmer *m*, Schnelleaufnehmer *m*	capteur *m* de la vélocité	snelheidsopnemer
V 16	velocity potential	Schnellepotential *n*	potentiel *m* de la vélocité	snelheidspotentiaal
V 17	velocity shock	Schnellesprung *m*	saut *m* de la vélocité	sprong in de snelheidsverdeling
V 18	velocity transducer	Schnelleaufnehmer *m*	capteur *m* de la vélocité	snelheidstransducent, drukgradiënttransducent
V 19	vented baffle	belüftete Reflexionsplatte *f*	baffle *m* aéré	schermplaat met gaatjes
V 20	vibraharp	Vibraphon *n*	vibraphone *m*	vibrafoon
V 21	vibrant sound	Vibrant *m*	son *m* vibrant, vibrato *m*	vibrerend geluid *n*
V 22	vibrate	schwingen	vibrer	trillen
V 23	vibrating chord	schwingende Saite *f*	corde *f* vibrante	trillende snaar
V 24	vibrating foil	schwingfähige (schwingende) Folie *f*	feuillet *m* vibrant, pellicule *f* vibrante	membraan *n*
V 25	vibrating pin	Membranstift *m*, Schwingstift *m*	tige *f* vibrante	trilpen
V 26	vibrating-string accelerometer	Saitenbeschleunigungsmesser *m*	accéléromètre *m* à corde	versnellingsopnemer met trillende snaar
V 27	vibration	Schwingung *f*, Vibration *f*	vibration *f*	trilling
V 28	vibration absorption	Schwingungsdämpfung *f*	amortissement *m* de la vibration	trillingsdemping
V 29	vibration acceleration	Schwingungsbeschleunigung *f*	accélération *f* de la vibration	trillingsversnelling
V 30	vibrational degree of freedom	Schwingungsfreiheitsgrad *m*	degré *m* de la liberté vibratoire	vrijheidsgraad voor de trilling
V 31	vibration amplitude	Schwingungsamplitude *f*	amplitude *f* de la vibration	trillingsamplitude
V 32	vibration displacement	Schwingungsausschlag *m*	élongation *f* de la vibration	trillingsweglengte

V 33	vibration frequency	Schwingungsfrequenz f	fréquence f de vibration	trillingsfrequentie
V 34	vibration isolator	Schwingungsisolator m, Körperschallisolator m	isolateur m de vibrations	trillingsisolator
V 35	vibration measurement	Schwingungsmessung f, Körperschallmessung f	mesure f de vibration	trillingsmeting
V 36	vibration meter, vibrometer	Schwingungsmesser m	vibromètre m	trillingsmeter, vibrometer
V 37	vibration node	Schwingungsknoten m	nœud m de vibration	trillingsknoop
V 38	vibration pick-up	Körperschallmikrofon n	capteur m de vibration	trillingsopnemer
V 39	vibration velocity	Schwingungsschnelle f	vélocité f de vibration	trillingssnelheid
V 40	vibrato	Vibrato n	vibrato m	vibrato
V 41	vibrograph vibrometer	Schwingungsschreiber m s. V 36	vibrographe m	trillingsschrijver, vibrograaf
V 42	vicinity field	Nachbarfeld n <einer Schallquelle>	champ m adjacent, champ m de voisinage	aangrenzend veld n
V 43	vielle	Radleier f	vielle f	lier, draailier
V 44	viol	Viole f	viole f	viola
V 45	viola	Bratsche f	violon m alto	altviool
V 46	violin	Violine f, Geige f	violon m	viool
V 47	violincello	Violoncello n	violoncelle m	violoncel
V 48	virginal	Virginal n <Spinettart>	virginal m <petit clavecin>	virginaal n
V 49	virginal groove	unmodulierte (unbeschriebene) Rille f <Schallplatte>	sillon m vierge	ongemoduleerde groef, maagdelijke groef
V 50	virginal tape	Leerband n <Tonband>	bande f vierge	maagdelijke band, onbespeelde band
V 51	viscous damping	viskose Dämpfung f, Flüssigkeitsdämpfung f	amortissement m visqueux	viskeuse demping
V 52	visible speech recording	frequenzspektrale Lautaufzeichnung f	enregistrement m spectral de la parole	zichtbare weergave van het spraakfrequentiespectrum
	visual reading	s. L 53		
V 53	vocal cavity	lautbildender Hohlraum m	cavité f vocale	stemholte
V 54	vocal consonant, voiced (sonant) consonant	stimmhafter Konsonant m	consonne f vocale (sonore)	stemhebbende medeklinker
V 55	vocal cord	Stimmband n	corde f vocale	stemband
V 56	vocal entry	Stimmeinsatz m	attaque f vocale	stem-aanhef
	vocal horn	s. F 52		
V 57	vocal range	Stimmumfang m, Stimmbereich m	étendue f d'une voix	stemspanwijdte, stembereik n
V 58	vocal tract	Rachenraum m	cavité f laryngienne	stemkanaal n
	vodas	s. V 67		
V 59	voder, voice operation demonstrator	Sprachsimulator m	simulateur m de parole	spraaksimulator
	vogad	s. V 68		
V 60	voice	Stimme f	voix f	stem
V 61	voice	stimmen <Musik>	accorder	stemmen <orgel>
V 62	voice-actuated	sprachbetätigt	commandé par la voix	op spraak reagerend
V 63	voice-actuated modulator	Sprachmodulator m	modulateur m de parole	door spraak ingeschakelde modulator
	voice coil	s. L 105		
	voiced consonant	s. V 54		
V 64	voice-ear test	Sprech-Hör-Versuch m	test m voix-ouïe	spraak- en gehoortest
V 65	voice frequency	Sprachfrequenz f	fréquence f vocale	stemfrequentie
V 66	voice-modulated	sprachmoduliert	modulé en fréquences vocales, modulé par la parole	met spraak gemoduleerd
V 67	voice-operated device antisinging, vodas	Vodas n	vodas m	vodas
V 68	voice-operated gain adjusting device, vogad	Vogad n	vogad m	vogad
	voice operation demonstrator	s. V 59		
V 69	voice power	Sprachleistung f	puissance f vocale	stemvermogen
V 70	voice reproduction	Sprachwiedergabe f	reproduction f de la parole	spraakweergave
V 71	voice transmission	Sprachübertragung f	transmission f de parole	spraak-overdracht
V 72	voicing	Stimmen n <Musik>	accordement m	stemmen
V 73	voix celeste	schwebende Stimme f <Musik>	voix f céleste	vox celesta <orgel>
V 74	voltage level gain	Spannungsverstärkungsmaß n	niveau m de gain en tension	spanningspeilversterking
V 75	volume	Lautstärke f	volume m (puissance f) sonore	volume
V 76	volume compression	Dynamikkompression f	compression f de dynamique	dynamiekcompressie
V 77	volume indicator	Aussteuerungsmesser m	indicateur m de puissance	volume-indicator
V 78	volume level	Sprachpegel m	niveau m de puissance sonore	volumepeil n
V 79	volume range	Lautstärkebereich m	gamme f de puissance	volumespanwijdte, volumebereik n
V 80	volume scattering coefficient	Volumenstreukoeffizient m	coefficient m de dispersion en puissance	volume-verstrooiingscoëfficiënt
V 81	volume unit, vu	Volum[en]einheit f, vu	unité f de volume, vu	volume-eenheid
V 82	volume velocity [across a surface element]	Schallfluß m [durch ein Oberflächenelement]	vélocité f de flux [à travers un élément de surface]	volume-snelheid [door een oppervlakte-element]
V 83	vortex	Wirbel m	tourbillon m	draaikolk, werveling
V 84	vortex wave	Wirbelwelle f	onde f tourbillonnaire	wervelgolf
V 85	vowel	Vokal m	voyelle f	klinker
V 86	vowel articulation	Vokaldeutlichkeit f, Vokalhervorhebung f, Vokalbetonung f	articulation f de voyelle	verstaanbaarheid van klinkers
	vu	s. V 81		

W

	wafer loudspeaker	s. F 43		
W 1	wainscoting	Täfelung f	lambris m	lambrizering, beschot n
W 2	wall admittance	Wandadmittanz f	admittance f de paroi	wand-admittantie
W 3	wall impedance	Wandimpedanz f	impédance f de paroi	wand-impedantie
	wand band amplifier	s. B 179		
W 4	wanted carrier	Nutzträger m	porteuse f utile	gewenste draaggolf
W 5	warble	wobbeln	vobuler	janken, kwelen <frequentieverandering>
W 6	warble frequency	Wobblerfrequenz f, Wobbelfrequenz f	fréquence f de vobulation	jankfrequentie
W 7	warble tone	Wobbelton m	son m de vobulation	janktoon, wobbeltoon
W 8	wave analysis	Wellenanalyse f, Frequenzanalyse f	analyse f d'une onde en fréquence	frequentie-analyse
W 9	wave angle	Strahlungswinkel m	angle m de radiation	golfhoek
W 10	wave concentration	Wellenbündelung f	concentration f d'onde, focalisation f	bundeling van golven
	wave crest	s. P 44		
W 11	wave echo	Wellenecho n	écho m d'onde	weerkaatste golf
W 12	wave filter	Wellenfilter n	filtre m d'onde	golffilter n
W 13	wave front	Wellenfront f	front m d'onde	golffront n
W 14	wave interference	Welleninterferenz f	interférence f d'ondes	interferentie van golven
W 15	wavelength	Wellenlänge f	longueur f d'onde	golflengte
W 16	wavelength coefficient	Phasenkoeffizient m	déphasage m caractéristique de longueur d'onde	golfgetal n
W 17	wavelength extension	Wellenverlängerung f	extension f de longueur d'onde	vergroting van de golflengte
W 18	wavelength shortening	Wellenverkürzung f	réduction f de longueur d'onde	verkleining van de golflengte
W 19	wave phase	Wellenphase f	phase f d'onde	fase van de golf
W 20	wave shape	Wellenform f	forme f d'onde	golfvorm
W 21	wave trap	Wellenfalle f	piège m à ondes	zeefkring
	wave trough	s. H 79		
W 22	wave velocity	Ausbreitungsgeschwindigkeit f	vitesse f de propagation	voortplantingssnelheid van de golf
W 23	weak	undeutlich <Sprache>, schwach <Signal>	faible, indistinct	zwak
W 24	weaken	abschwächen, schwächen	affaiblir	verzwakken
W 25	Weber-Fechner law of hearing	Weber-Fechnersches Gesetz n	loi f de Weber-Fechner	wet van Weber-Fechner voor het gehoor
W 26	weight	bewerten	évaluer	wegen
W 27	weighted sound pressure level	bewerteter Schalldruckpegel m	niveau m de pression sonore évalué	gewogen geluiddrukpeil n
W 28	weighting	Bewertung f	évaluation f	weging
W 29	weighting filter	Bewertungsfilter n	filtre m d'évaluation	weegfilter n
W 30	weighting network	Bewertungsschaltung f	réseau m d'évaluation	weegnetwerk n
W 31	whispered speech	Flüstersprache f	voix f basse, murmure m	gefluister n
W 32	whistling tone	Pfeifton m	sifflement m	fluittoon
	white noise	s. U 37		
W 33	whiz, hiss	zischen	chuinter, siffler	sissen, fluiten
W 34	whole-number harmonic	ganzzahlige Harmonische f	harmonique f entière	geheeltallige harmonische
W 35	wide-band artificial ear	künstliches Ohr n mit Breitbandcharakteristik	oreille f artificielle à large bande	breedbandige oorsimulator
W 36	wide-band microphone	Breitbandmikrofon n	microphone m à bande large	breedbandige microfoon
W 37	wide bandpass filter	Breitbandfilter n	filtre m à large bande	breed bandfilter
W 38	wind noise	Windgeräusch n, Störschall m <Mikrofon>	bruit m de souffle	windgeruis n
W 39	wind trunk	Windkanal m <Orgel>	laye f	windkanaal <orgel>
W 40	wind valve	Spielventil n <Orgel>	soupape f	speelventiel n, cancelventiel n <orgel>
W 41	wobbling frequency	Heulfrequenz f, Wobbelfrequenz f	fréquence f de vobulation	jankfrequentie, wobbelfrequentie
W 42	woodwind instrument	Holzblasinstrument n	bois m <instrument de musique>	houtblazersinstrument n, houten blaasinstrument n
	woofer	s. B 159		
W 43	wooliness	starker Nachhall m	écho m puissant	te sterke nagalm
W 44	working wave	Betriebswelle f	onde f de travail	werkzame golf
W 45	wow	Jaulen m	hurlement m	lage jank
W 46	wow	Schwankung f der Bandgeschwindigkeit	fluctuation f lente de la vitesse de transport	ongelijkmatige snelheid <van band of plaat>
W 47	wrest	Stimmeisen n	clé f d'accordeur	stemsleutel
W 48	wrest block <plank>	Stimmstock m		stemblok n

X

	X-mit	s. T 127		
	X-mitter	s. T 130		
	X-mitting	s. T 132		
X 1	X-stopper <US>	Störschutz m gegen atmosphärische Störungen	suppresseur m de parasites atmosphériques	storingsonderdrukker
X 2	xylophone	Xylophon n	xylophone m	xylofoon
X 3	X-Y technique	XY-Schallaufnahmeverfahren n <Stereotechnik>	enregistrement m stéréophonique XY	X-Y-opname <stereofonisch>

Y

Y 1	yelling	gellend	strident, criant	gillend
Y 2	yieldingness	[akustische] Nachgiebigkeit f	flexibilité f, souplesse f	slapheid

5*

Z

Z 1	**Zener noise**	Zener-Rauschen *n*	bruit *m* de Zener	Zener-ruis
Z 2	**zero beat**	Schwebungslücke *f*	trou *m* de battement	zwevingsfrequentie nul
Z 3	**zero-beat reception**	Homodynempfang *m*	réception *f* homodyne	homodyneontvangst
Z 4	**zero frequency**	Nullfrequenz *f*	fréquence *f* zéro	frequentie nul
Z 5	**zero insertion loss**	Nulldurchgangsdämpfung *f*	pertes *fpl* de passage de zéro	geen tussenschakeldemping
Z 6	**zero-order radiator**	Kugelstrahler *m* nullter Ordnung	source *f* omnidirectionnelle d'ordre zéro	geluidbron van de nulde orde, puntbron
Z 7	**zero output**	Nullsignal *n*, Ausgangswert *m* Null	puissance *f* de sortie nulle	geen uitgangsvermogen *n*
Z 8	**zero-phase modulation**	Nullphasenmodulation *f*	modulation *f* de phase zéro	modulatie zonder fase-verschuiving
Z 9	**zone of dispersion**	Bereich *m* der Dispersion, Dispersionsbereich *m*	zone *f* de dispersion	dispersie-gebied *n*

DEUTSCH

FRANÇAIS

NEDERLANDS

basreflexkast B 72, R 93
bassethoorn A 192
bastrombone B 76
bathythermogram B 79
becijferde bas F 26
bedekken C 93
beeldomvormer I 4
beenviool B 77
begeleidende muziek I 31
beginstrook L 15
begrenzen van de topwaarden P 38
begrenzende straal L 37
begrenzende versterker L 36
begrenzer C 83
begrenzing C 84
begrenzing van de frequentie-band L 34
begrijpelijk R 34
begrijpelijkheid R 33
beide oren, voor B 116
beitel E 55
beker B 102
bekken C 251
bekkens C 252
beklede geluidband C 96
beklemtonen A 21
beklemtoning A 22
beklemtoning, te sterke — O 42
bel B 101, B 104
belastbaarheid P 151
belastingsimpedantie L 67
bellen D 105
benadering van een grenslaag B 167
beoordeling van de luidheid L 92
bepaling van het geruisniveau N 76
beroepsmatige lawaaidosis O 4
beschikbaar ruisvermogen A 330
beschikbaar vermogen A 331
beschot W 1
bestralende geluidpeil I 145
bestraling met geluid S 204
beveiliging tegen ongewild wissen E 86
beweeglijke microfoon F 59
bewegingsimpedantie M 142
bilabiaal B 120
bilateraal hoortoestel B 122
bilaterale microfoon B 124
binauraal hoortoestel B 128
binnenkomende golf I 35
blaasorkest B 171
blikkerig geluid T 63
blokfluit B 141
blokkeerkarakteristiek B 143
blokkeerverzwakking B 142
blokkeren B 138
bolbron S 283
bolvormige golf S 284
bombardon B 153
borststem C 50
botgeleiding B 154
botgeleidingsmicrofoon B 156, O 34
botgeleidingstelefoon B 155, O 32
botgeleidingstriller B 157
bouwakoestiek A 232
boven de gehoorgrens S 376
boventoon O 57, U 51
bovenste begrenzing H 66
bovenste grensfrequentie H 64
breed bandfilter W 37
breedbandige microfoon W 36
breedbandige oorsimulator W 35
breedbandige ruis B 182
breedbandige symmetrische transformator B 180
breedbandige versterker B 179
breedbandig filter B 181
breedbandversterker A 187
breking R 96
brekingsas A 338
brekingsverlies R 97
brom H 95

brom, het oppikken van — H 105
bromcomponent H 98
bromfactor S 92
bromfiltering H 100
bromfrequentie H 101
bromhout B 197
brommen H 94
brommen tijdens onder-titeling C 11
bromniveau H 96
bromstoornis H 107
bron van lawaai N 71
bronimpedantie S 255
bronsterkte op de as A 337
bronsterkte op de as ‹transdu-cent› A 337
bronsterkte van de sonar S 137
brug-evenwicht B 176
brugschakeling L 13
brugtak B 175
bruikbare veld U 53
brulboei A 55
brullen B 107
Buchmann-Meyerpatroon B 189
bufferversterker B 190
buigen B 89
buiggolf B 112, F 49
buiging D 91
buigingsgebied D 95
buigingspatroon D 93
buigingsrooster M 168
buigingsverlies D 94
buigtriller F 48
buigzaam membraan F 46
buik L 83
buis van Eustachius A 308, E 88
buis van Kundt K 9
buisbrom V 1
buisruis V 2
buitenzintuigelijke waarne-ming E 104
bulderen B 158
bulderend B 163
bulderend geluid B 164
bundel afgebogen golven D 92
bundelbreedte B 83
bundeldoorsnede C 236
bundelen C 146
bundeling B 81
bundeling van geluid A 61
bundeling van geluidver-mogen A 117
bundeling van golven W 10
buurkanaaldemping S 28
bijgeluid I 108
bijstemmen R 152
bijtonen A 162
bundelingsfactor D 113, D 124
bijtoon S 24, S 89

C

cancelventiel ‹orgel› W 40
capacitieve pick-up C 7
cardioïdmicrofoon C 15
carillon C 16
castagnetten C 22
cavitatie C 24
cavitatiebel C 25
cavitatiegeruis C 26
cavitatiekern C 27
celesta C 30
cembalo C 31
cent C 34
centibel C 33
centiem C 34
centrale studio M 53
centrale versterkingsregeling M 52
centreerspin I 67
chinese bekkens C 52
chirurgie met geluid S 151
chromatische toonladder C 63
chromatisch interval C 62
cilindrische golf C 250
cimbaal C 251
cinema-geluidweergave-systeem M 143
circumaurale telefoon C 70
citer C 72

clavecimbel H 29
clavichord C 79
coaxiale luidspreker C 97
coaxiale verzwakker C 98
coëfficiënt van akoestische dissipatie S 183
coëfficiënt van faseverschui-ving P 74
coïncidentieversterker C 105/6
combinatiepedaal D 183
combinatietoon C 111
combinatietoonvervorming C 112
commando-microfoon I 103
compensatie van filterverzwak-king C 122
compenseren door tegenkop-peling B 115
complexe aanstoting C 129
complexe admittantie C 128
complexe grootheid C 130
complexe impedantie C 133
complexe responsie C 131
compliantie C 135
compressor C 145
concertina C 150
concertvleugel C 148
concrete muziek C 153
condensatorluidspreker C 8
condensatormicrofoon C 9
condensatortelefoon C 10
conductantie C 156
conductantie voor ruis N 40
conische hoorn C 161
conservatieve flux C 163
constantheid van de toonhoog-te C 165
constructiegeluid S 343
constructiegeluidfilter S 344
contactgeluid I 15
contactgeluidpeil I 16
contactgeluidspectrum S 266
contactmicrofoon C 169
continu C 171
continu geluid C 174
continu geluidpeil C 175
continu spectrum C 176
continu systeem C 178
contrabas B 71, C 179
contrabasklarinet C 180
contrabastrombone C 182
contrafagot C 181
contrapunt C 205
contrapuntisch C 185
contrastregeling A 166
controle van het richteffect S 59
controleluidspreker M 135
controleren M 132
controlespoor C 190
controletelefoon M 133
controleversterker M 134
conusluidspreker C 159
conusluidspreker voor lage tonen L 116
conusmembraan C 158a
convergentiezone C 191
convergerende golf C 192
conversiegrafieken voor tril-lingspatronen C 193
correctie voor de hoge tonen T 141
correctie voor lage tonen B 62
correctienetwerk voor opname R 62
correctiespoel A 221
Coulomb-demping C 201

D

dalend geluidpeil D 59
darmsnaar G 39
-3 dB-punt H 2
decibel D 31
decorluidspreker B 2
deel van een trede F 70
deeltjessnelheid P 23, S 251
deeltjesverplaatsing P 22, S 217
deeltoon P 21
delerketen D 160
delernetwerk D 161
demodulatie D 54

demodulatieschakeling D 56
demodulatietrap D 55
demodulator D 63
dempen D 1a, D 15, M 156, M 185
demper A 6, D 6, M 158
demping D 8
demping door luchtwrijving A 181
demping in een kamer R 182
demping van luchttrillingen D 10
dempingsconstante A 254
dempingscorrectie A 255
dempingsfactor D 9, D 11
dempingsmoment D 12
densitometer D 57
derdegraads vervorming T 39
detectie D 54
detectiedrempel D 61
detector D 63
determinant van een netwerk N 23
diatonisch D 76
diatonische accordeon P 202
diatonische toonladder G 6
diatonisch interval D 77
diatonisch systeem D 78
diepe modulatie H 50
diepe verstrooiende laag D 39
dieptechrift H 76
differentiaaldrempel van her-kenning R 51
differentiaalkoolmicrofoon D 169
differentiaalmicrofoon D 86
differentiëerschakeling D 88
diffuse inval R 14
diffuse terugkaatsing D 99
diffusiteit D 102
diffuus geluid D 100
diffuus veld D 96
diffuus verstrooiend oppervlak D 101
dilatatiegolf D 103
dip D 106
directe registratie D 126
directe uitzending O 17
discant D 58
discantluidspreker T 165
discriminatieverlies D 129
discriminator F 83
discrimineren tegen harmoni-schen D 128
disharmonie D 132
disharmonisch D 131
dispersie D 136
dispersie-gebied Z 9
dissipatie D 140
dissipatiefactor D 141
dissonant D 127
distorsie D 148
dubbele geluidbronlocalisator T 166
dubbelsporige opname T 167
doedelzak B 16
doedelzakspeler P 113
doelsterkte B 10
dof B 144, D 206
dofheid D 207
dokje ‹clavecimbel› J 1
dominant akkoord D 163
dominante frequentie D 164
doof D 19
doofheid A 206, D 23
doofstom D 21
doofstomheid D 22
doordringen P 55
doordrukeffect A 23, M 14, P 172, S 294
door-en-door magnetische band D 134
doorgaande golf T 54
doorgelaten golf T 129
doorlaatbaarheid T 120, T 128
doorlaatbaarheid voor geluid S 219
doorlaatbaarheidsfactor T 124
doorlaatbaarheidskarakteris-tiek T 123
doorlaatband P 25, T 121
dóórschieten O 55
dopplereffect D 165